Unnatural Landscapes

Unnatural Landscapes

Tracking Invasive Species

Ceiridwen Terrill | Foreword by **Gary Paul Nabhan**

The University of Arizona Press   Tucson

Map illustrations are by Rick Moser.

All photos were taken by the author unless otherwise indicated.

The University of Arizona Press
© 2007 The Arizona Board of Regents
All rights reserved

Library of Congress Cataloging-in-Publication Data
Terrill, Ceiridwen, 1970–
    Unnatural landscapes : tracking invasive species /
Ceiridwen Terrill ; foreword by Gary Paul Nabhan.
        p. cm.
    Includes bibliographical references and index.
    ISBN-13: 978-0-8165-2523-2 (pbk. : alk. paper)
    ISBN-10: 0-8165-2523-4 (pbk. : alk. paper)
    1. Introduced organisms—North America. 2. Island
ecology—North America. I. Title.
QH102.T47 2007
577.5'218—dc22
2006019516

Publication of this book is made possible in part by the pro-
ceeds of a permanent endowment created with the assistance
of a Challenge Grant from the National Endowment for the
Humanities, a federal agency.

Manufactured in the United States of America on acid-free,
archival-quality paper.

12 11 10 09 08 07   6 5 4 3 2 1

Contents

Foreword   Gary Paul Nabhan

It may be like seeing the world we live in freshly, as if for the first time. Ceiridwen Terrill brings us to scenes that some of us may live near, walk past, or drive by many days of our lives, but they are scenes that have gone unnoticed, struggles between natives and invasives that we have rarely glimpsed. Like it or not, we are part of an era that later generations will describe as one preoccupied with homeland security, and yet the very biological issues that are truly destabilizing and impoverishing our homelands and the waters within them are barely addressed.

Ceiridwen Terrill seeks to correct that myopia not as an optometrist might, but through telling us stories. As we listen, we hear what we have not quite ever imagined, and our eyes grow wide. She offers us parables from island life, which may remind us that the very same processes that can easily be circumscribed within such islands are happening in a more diffuse manner all around us. Yes, it may be true that on islands there is a peculiar distortion and magnification of the effects of exotic species on their competition with natives. But, after all, that is exactly what a parable is: an elegant microcosm that heightens our understanding of a more general phenomenon by making it poignantly memorable.

I have been on some of the very islands that Ceiridwen Terrill portrays, and I recognize that, to the uninitiated, the dramas unfolding there are sometimes touched with the surreal. You may arrive safely on a shore only to face an impenetrable mass of thousands of tamarisks and giant *carrizo* canes, which you must climb over or through just to reach the remnant habitat of an endemic pack rat or wildflower. Like being on the islands dubbed "Las Encantadas" in *A Midsummer Night's Dream*, you may have the odd feeling that everything around you is a momentary mirage, a fleeting presence on the face of the Earth. If the continued dispersal and competition of invasive species is not in any way checked, what we see everywhere in the deserts of the American Southwest and Baja California may be a fleeting reality. Bring a Hawaiian born four centuries ago back to the Big Island today through some time machine, and he or she would not recognize it as home. Our globalized, technological culture may not be able to successfully build time machines yet, but we are fantastic at telescoping the time that it once took to dramatically alter some island from four centuries down to four decades or four years.

But let me be clear about two aspects of this book. First, Ceiridwen Terrill is not attempting to recapture some pristine past in which the West or Island X was "100 percent natural," as if cultural influences have not continued to shape and reshape the landscapes within which we live. Here she is offering a deep perspective on the *dynamics* of change and she is often focused on the complex human responses to it. Some of the protagonists in this book are my heroes, as they are hers, but all of them struggle to offer the appropriate ethical and emotional responses to such cataclysmic ecological changes that are occurring in our midst. Second, this is not the contrived reporting of "front-line journalists" embedded with the troops, who see only what the propaganda minister wishes them to see. Ceiridwen is a good naturalist and a broad thinker who challenges some of the assumptions made by the scientists fighting against certain weedy invaders. In that sense, this is a critical look at the state of the art of controlling invasives in the context of our larger society, rather than simple reportage.

Yet, I suspect that like good cheese or wine, this book will gain value with age, because it offers a glimpse of a critical moment in the era that David Quammen has called "the Homogecene," an epoch of anthropogenically triggered declines in diversity. It is a critical moment of self-reflection by the species that caused this global homogenization. We can choose to continue to let each distinctive place on Earth become more like every other, or we can enter "the Heterogocene," an era in which diversity and distinctiveness are once again protected, respected, celebrated, and restored. It takes a distinctive vision to elucidate the way out of our current mess; Ceiridwen Terrill provides us such a guiding light.

Gary Paul Nabhan

Unnatural Landscapes

Introduction The Voyage

If I stopped and thought, maybe
the world
can't be saved,
the pain was unbearable.
—Mary Oliver, from the poem "The Moths"

From the rear cockpit of my sea kayak, *The Grebe*, I paddle solo, cicum-
navigating Anaho Island in Pyramid Lake, Nevada, where a year ago I
helped count the island's nesting birds with the U.S. Fish and Wildlife

The pyramid of Pyramid Lake, Nevada

Service. This year's nesting season is nearly over now. I brace against September winds tumbling off the jagged peaks surrounding the lake and with rhythmic paddle strokes sweep dark water to the stern of the boat. This is a tandem kayak, meant for two people paddling in sync, the person in the rear pressing foot pedals to move the rudder from side to side to steer. Working as a team and using abdominal muscles to avoid overdependence on the arms, tandem kayakers conserve energy during long-distance crossings between islands, which is what my husband Bruce and I have been doing for the last year as I prepared to write this book.

Anaho Island is the first of several islands I've explored with the aim of alerting readers to the conservation problems caused by exotic plant and animal species. Exotics are species introduced to an area where they don't occur naturally and where they may become invasive, threatening the health of native species and even whole ecosystems.

As I paddle, I recall a conversation with a historian friend of mine. "I'm all for history," he said. "As far as those invasive species you're talking about, one of the reasons to study the history of their introductions is to figure out what the mistakes were and then do your best to fix them." I agreed with him and pointed out that, while many of these introductions occurred within the last century or even earlier, some are recent, and introductions are still occurring today. My hope for this book is that it will make regular people—the millions of us who visit sensitive habitats each year—want to learn about mistakes that may have been made in our own regions and help to repair damage and avoid invasive-species introductions in the future.

Islands are excellent places for exploring the problem of invasive species because their native plants and animals are highly specialized, isolated organisms, often few in number, and highly susceptible to the negative effects of introduced species. According to biologist Bernie Tershy, executive director of Island Conservation, these qualities make islands ideal laboratories for learning about how nonnative plants and animal species travel, establish, and spread.

Tershy explains that invasive species tend toward higher densities on islands because their natural predators or herbivores are not present to keep their numbers in check. Also, many exotics have resistance to ecological pressures that native species must contend with, such as climate change and disease. Native populations also tend to be small on islands, so invaders can more easily overtake them. Tershy points out that, although islands make up only 3 percent of the Earth's surface, they are inhabited by 15 percent of the world's plant, bird, and reptile species. Since the 1600s, 80–90 percent of the Earth's bird and reptile extinctions, as well as 50 percent of all plant and mammal extinctions, have been of island species. But he emphasizes that islands offer great opportunities for conservation work, "Because islands have such strict perimeters, with the right resources you can completely eradicate an invader from an island with less risk of reinvasion." Islands provide quick feedback about whether a particular eradication technique is working. If an invader is still present after an eradication effort, it can be more easily detected on an island than on the mainland. "When you eliminate an invasive species from an island," Tershy says, "you see dramatic comebacks by native species such as seabirds that nest exclusively on islands."

According to Tershy, "Islands are central to conserving biodiversity worldwide," so I've chosen four sets of islands to focus on: Nevada's Anaho Island; Ash Meadows National Wildlife Refuge and its California neighbors Death Valley and Fish Slough, whose small pools and springs surrounded by desert reverse our usual notion of islands; the Midriff Islands in the Sea of Cortés, Mexico; and California's northern Channel Islands.

1   Anaho Island
    Pyramid Lake, Nevada

2   Ash Meadows, Nevada
    Death Valley, California
    Fish Slough, California

3   Midriff Islands
    Sea of Cortés, Mexico

4   Anacapa and Santa Cruz Islands
    Channel Islands, California

Map of the island habitats discussed in the book. Illustration by Rick Moser

Because both plants and animals can be invasive, two of the book's chapters are dedicated to plant invaders and two explore the impacts of invasive animals. All these islands are at different stages of invasion by exotic species, and all are accessible to readers. Even Anaho, normally off limits to the general public, can be visited by volunteering to count nesting birds, as I did.

All the islands in this book present unique problems to conservationists who want to restore their native ecosystems, and my job has been to research some of those challenges. I've learned that invasive species haven't just outcompeted natives for nutrients, space, and water. They can also change fire regimes, alter stream courses, prevent native plants from regenerating, negatively affect human health and local economies, and, in some cases, become the subject of intense emotional debate in nearby communities.

As I paddle along the western side of Anaho I encounter a group of Clark's grebes bobbing on the water. They curl their willowy dark necks in a diver's tuck and thrust their bodies below the surface to catch fish, emerging moments later to look around coolly, at ease with their dives into the dark. Our kayak is named after this bird because its sleek design and Greenland bow allow it to tunnel through large waves and burst out on the other side, seawater sheeting off its deck. A silent craft, the kayak enables Bruce and me to approach animals without disturbing them and to venture into places where larger boats can't go.

To become as confident as the grebes for our trips to the islands, Bruce and I paddled Pyramid Lake, strengthening our abdominal muscles and practicing wet exits. During these rehearsals, we purposely capsized the kayak, going through the drill of counting to three while hanging upside down underwater, a technique for calming the mind and preventing panic. After three counts, we pulled the safety loops on our spray-skirts and wriggled free of our cockpits, then swam to the surface.

Alone on the lake today, I round the northern tip of Anaho, careful to maintain the legal distance of 500 feet from the island's shoreline. The small Paiute village of Sutcliffe, situated on the lake's western shore, disappears from sight, and I watch pelicans feed nearby, dipping their bills in tandem as they fish. Beyond them, on the island, I can make out the blonde stems of cheatgrass and the reddish hue of red brome, both invasive grasses introduced accidentally to the island.

Rhythmic paddle strokes propel my thoughts forward, and I go over in my mind other ways invasive species get to places they don't belong. Intentional introductions rank high on the list. In California, for example, several county health departments deliver invasive mosquito fish free of charge to anyone who wants them. This fish, measuring two to three inches long and resembling a guppy, is touted as an effective mosquito-control agent that may help prevent the spread of West Nile virus by eating mosquito larvae. But mosquito fish are overrated as mosquito-control agents, because they eat fewer larvae than do native fish. These invaders have overrun the fragile reverse islands of Nevada and California, where they eat the eggs and young of native fish and compete with them for limited food resources. The mosquito fish has also pushed the California newt and the California tree frog toward endangered status.

Human feelings sometimes keep us from making the health of native species a top priority and lead to intentional introductions of exotics. During my volunteer work with the Santa Barbara Wildlife Care Network, the manager of a local hardware store stumbled into the center holding a glue tray with two baby black rats stuck by their chins, feet, and tails to the adhesive, awaiting death by starvation. The wildlife-care director pulled a jar of mayonnaise from the refrigerator and worked gently with this "miracle cream" for removing sticky substances from wildlife. She massaged it between the rats' toes and into their fur, easing their bellies, chins, and tails from the tray.

Although I abhorred the use of the glue tray, I suggested that the rats should be humanely euthanized. A fellow volunteer responded with horror and disgust, "How can you say such an awful thing? Why would you kill them?"

"These rats are invasive," I said. "They prey on rare seabirds."

"I don't care. They deserve to live like any other animal."

The decision was out of my control. Rehabilitation and release was the policy of the center, and it extended to all species, whether native or exotic. The plan for our rat patients was rehydration, food, and expensive antibiotics. When they were well, they'd be released. The rats peered from their den of tissue paper with inquisitive black eyes, taking in the world that had nearly killed them, and I watched while precious donations of money and supplies that could have been used to rehabilitate injured native animals were poured into saving the lives of these invaders.

Exotic pets are another source of introductions, both intentional and accidental. A quick glance at the classified section of any local newspaper yields advertisements for ferrets, iguanas, red-tail boas, baby bearded dragons, pot-bellied pigs, even hedgehogs. Escaped pets, or those intentionally released into the wild when owners can no longer care for them, can do a lot of damage to native wildlife. Even when a pet is secure with its owner, it can introduce foreign parasites and diseases. Dogs, while likely immunized against canine distemper, can nevertheless be carriers, transmitting the disease to native island inhabitants.

People are largely unaware of the dangers of bringing their pets to islands. Take the example of the man who realized only when he received a customs-declaration form from a flight attendant that he was breaking Hawaiian law by bringing along his pet, a desert rosy boa native to California, one of four invasive snakes most frequently spotted loose in Hawaii's habitats and common to the pet trade. Fortunately, Hawaii has a well-organized system for keeping out invaders. As visitors to the islands deplane, they're confronted by amnesty bins, containers that offer them the chance to get rid of illegal biological material before being caught and fined. All sorts of exotics end up in those bins; once someone even dropped in a live ball python. Nevertheless, snakes are still smuggled into Hawaii as pets or they arrive in cargo shipments.

The increased liberalization of global trade, without consideration for potential introductions of invasive species, is a major source of accidental introduction. Invaders arrive in the ballast water of ships and in cargo, and the nursery trade transports millions of unseen organisms around the world in potting soil.

While transporting supplies and people all over the world, the military has also been a source of exotic species introductions. An accidental introduction on Guam created an island almost entirely without birdsong. In July 1944, at the height of World War II, the U.S. military drove the Japanese off this northern Pacific island and took it over. Brown tree snakes coiled in the wheel-wells of fighter planes, stowed away in cargo shipments and in the hulls of navy ships, and hitchhiked to Guam from their native homes in the Solomon Islands, New Guinea, eastern Indonesia, and northern and eastern Australia. Perhaps a single pregnant female founded the first island colony. But more likely several snakes were introduced over successive trips, found each other, and mated. Now for every square mile of island habitat in Guam, 10,000 to 13,000 brown tree snakes lurk about, causing frequent power outages as they climb telephone poles and transformers. Excellent climbers, brown tree snakes multiplied to their present numbers by feeding on the eggs and chicks of the island's native birds, driving several unique species to extinction.

Eradicating nonnative plants and animals is a messy business and fraught with difficult choices. Although a plant may be invasive, it may also provide critical habitat for native bees, butterflies, and birds that once relied on indigenous plant species no longer available to them. In southern California, for example, where people have converted anise swallowtail butterfly habitat to agricultural fields and grazing pastures, the butterfly's food sources, the native bladder parsnip and cow parsnip, have been largely replaced by exotic species. Luckily, this butterfly has been able to improvise with an invasive substitute, European sweet fennel—a small compensation for the loss of habitat. But not all native species can make do, and

An anise swallow-tail caterpillar makes do with invasive plant substitute.

sometimes those invasive substitutes bring with them other disadvantages. Sweet fennel, for example, infuses the soil with a chemical cocktail that inhibits the growth of native plant species.

Overhead, a single-file line of American white pelicans ripples over the top of Anaho. After several powerful strokes, I rest my arms and let the kayak glide while I watch the birds in flight. Energy conservationists, pelicans don't waste a single movement. They alternate between slow wing beats and coasting. Like long-distance cyclists, the lead bird sets the cadence and each pelican drafts off the bird ahead of it. In this way the birds can fly long distances without tiring.

This is the kind of stamina required for dealing with the problem of invasive species. Earth itself is an island no larger than a penny in the universe and cast with the same greenish hue as one of those old copper coins. Currently, it's the only place we can call home. Conservation biologists say we are living in the sixth great extinction event in Earth's history, one caused by our own hand and one we have the ability to slow or halt altogether. I'm optimistic that if people understand the significant threats invasive species pose, they will be eager to be part of the solution.

Chapter 1   Not Only for the Birds

Today the honey-colored hills that flank the northwestern mountains derive their hue not from the rich and useful bunchgrass and wheatgrass which once covered them, but from the inferior cheat. . . . The motorist who exclaims about the flowing contours that lead his eye upward to far summits is unaware of this substitution.

—Aldo Leopold, *A Sand County Almanac*

At the southern end of Pyramid Lake, Anaho Island looks like a horned lizard surfing. Thirty-five miles northeast of Reno, Nevada, "The Biggest

Little City in the World," this island of serrated rocky surfaces and spiky crags received wildlife-refuge status in 1913 because it harbors one of the largest nesting populations of the American white pelican in the United States. Although Anaho is located on lands of the Pyramid Lake Paiute Indian Reservation, the U.S. Fish and Wildlife Service holds jurisdiction over the refuge, and Donna Withers, with the Fish and Wildlife office in Fallon, Nevada, manages the island. The agency would likely pay more attention to other animal and plant life if it had the money. But it doesn't, so the staff funnels what little they do have into supervising and counting migrating native white pelicans.

Although the pelicans interest me, I want to understand threats to the health of Anaho's native plants and other animals, particularly the island's rattlesnake, which may be endemic. To see the island for myself, I've arranged to accompany Donna Withers and operations specialist Styron Bell as a volunteer who will help count pelicans on Anaho one morning a week for the next twelve weeks.

So I wait, prowling around the marina in Sutcliffe, a small Paiute community on the shores of Pyramid Lake. Loitering along the shoreline, I keep an eye out for Donna behind the wheel of a green U.S. Fish and Wildlife pickup. When the marina store finally opens, I clutch my coat against the chill and trot up the sandy bank for a cup of coffee.

During the 1960s, Verne Woodbury, then a graduate student in biology at the University of Nevada, Reno, conducted the only comprehensive study of Anaho Island's native and exotic plants and animals. Woodbury's

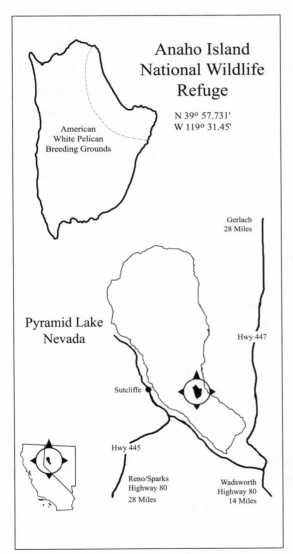

Anaho Island National Wildlife Refuge

N 39° 57.731'
W 119° 31.45'

American White Pelican Breeding Grounds

Gerlach 28 Miles

Pyramid Lake Nevada

Hwy 447

Sutcliffe

Hwy 445

Reno/Sparks Highway 80 28 Miles

Wadsworth Highway 80 14 Miles

Anaho Island National Wildlife Refuge in Pyramid Lake, Nevada. Illustration by Rick Moser thesis, completed in 1966, never sold in bookstores, although students and professionals frequently borrow the manuscript from the university library. Woodbury gathered the data for his thesis forty years ago, and, to date, no one has produced a follow-up study to ascertain changes to the island.

Plucking a cup from the dispenser in the marina store, I wonder how much the island's plant and animal community could have changed in forty years. Folgers dribbles from a pour spout on the shiny stainless steel percolator, the coffee bubbling into my Styrofoam cup. The kid running the cash register sees me peering at early photos of fishermen hanging on the wall of the store. One black-and-white shot shows a fly fisherman in hip waders casting his line for the famous cutthroat trout.

"You know, Clark Gable and Marilyn Monroe came here," the kid says.

I knew that during the filming of "The Misfits" (1961)—the last film that Clark Gable and Marilyn Monroe completed before their deaths—Gable's favorite pastime was fishing Pyramid Lake.

"The lake is nice, huh?" The boy looks about eighteen by the acne on his forehead. He tells me he lives on the reservation with his Paiute girlfriend. "Yeah, it's a nice lake, even if it is cursed. Least that's what they say out here. You gotta be careful. They say Indians float, but white men sink."

I ask what he means, and he says that more white people than Paiutes drown in Pyramid Lake. "It's a fact. You can look it up," he adds. According to the Paiutes, Indians float, but the Water Babies, lake spirits, pull the *Hyko* (white man) to the bottom of the lake so his body can't be found.

I'd read stories describing Water Babies as spirits of the Paiute infants bundled into baskets and launched into the lake when white settlers and soldiers attacked Paiute villages. The soldiers slaughtered Paiute mothers and fathers, old men and women. With no one to recover them, the infants drifted until they drowned. Their spirits haunt the lake, drowning white men for revenge.

Plunking down quarters for the coffee, I nod and slip out the door to sit on the dock next to the boat launch, where two pelicans engage in ritual courtship-feeding. The birds glide across the lake and rhythmically pierce the water's surface with their orange bills. Nuptial tubercles, horny knobs on top of each bill, indicate that these birds are mature enough to breed. These keel-shaped growths fall off once the parent birds begin feeding their chicks. The change in breeding status, from courtship and egg-laying to chick-rearing, triggers the loss of the knobs and the appearance of black feathers seated on the backs of the pelicans' heads like velvet yarmulkes.

Unlike its endangered cousin, the California brown pelican, the American white pelican is not listed as endangered in the state of Nevada, although it is listed as endangered in Washington State. In Nevada, money has been shunted away from consistent long-term studies to determine any potential threats to this bird. In fact, until the 1990s, if people wanted to know a bit of natural history about the white pelican, they had to consult literature on the brown pelican, even though the two species are radically different in size, feeding and nesting behaviors, flight patterns, and over-wintering habits, right down to the type of lice that live on their bodies.

It's like comparing lentils to peanuts. They're both legumes, but nothing alike. Despite the white pelican's radical differences from the better-studied species, there's little money for monitoring these birds closely, leaving U.S. Fish and Wildlife to rely on occasional volunteers like me to do the annual bird counts during nesting season.

I slurp my coffee, watching the birds hunt in tandem, totally alert for the slightest movement of fish beneath them. Like water-ballet performers in perfect unison, the birds plunge their heads and part of their necks underwater, trawling for fish. For efficiency, pelicans angle in perfect accord. If they randomly dip their bills, the fish will scatter, but by trawling in unified fashion, pelicans keep their prey in a school and each pelican in the group is more likely to trap a fish in its bill.

Typical of early spring in the high desert, the morning air is chilly, and by the time Styron and Donna pull up, I'm shivering inside my coat. Styron hops out of the truck to survey the sky to the west, over the Virginia Range, and shakes his head. "We better make it quick today. I don't like that sky." Then he glances over at Donna, his supervisor, to make sure he made the right call. Donna nods.

Bob, a U.S. Fish and Wildlife maintenance worker, unhitches the boat and backs the E-Z Loader down the ramp into the water. The boat slips in, and we pile our gear into a shallow cubbyhole at the front of the boat, where important equipment like cameras and spotting scopes won't get wet. As the April sun hangs at a forty-five degree angle to the island, we bundle for the windy crossing and strap on life vests.

We don't dally when Bob waves us on deck. He starts the engine and eases the boat away from the marina. He'll wait until we're about fifty yards from shore before putting the boat into high speed for the three-mile crossing to Anaho. The eastern sky over the island is blue and streaked with cirrus clouds, but weather coming from the west, from the Pacific Ocean, displays a bilious green color.

"What we do out here depends on the weather, and Pyramid is all weather," says Donna, blowing her nose. "It's unpredictable out here."

Some unwitting bird-counting crews have stayed the night on Anaho's windy shores, bunking down behind shrubs and boulders to get out of the wind. Weather creeps over the mountains and strands them on the island before they recognize the change. Without benefit of food or warm clothing, many a glum researcher has experienced an impromptu slumber party surrounded by dusty bird bones and brackish water. Anaho Island has no fresh water, and Pyramid Lake is approximately one-sixth as salty as the ocean, so, although it's technically a freshwater lake, you wouldn't want to drink from it.

Recreational boaters also get stuck on the island when a stalled engine leaves them vulnerable to wind and current. Others have capsized and drowned, their bodies never found, rekindling the story of the Water Babies. Locals, both white fishermen and Paiutes, have advised me to watch the sky to the north and west for even a hint of strong winds or clouds simmering over the mountains. If I see those clouds, they said, get off the water.

With inclement weather and an insufficient number of well-trained, paid bird counters, data collection on the white pelican project has been catch as catch can. For manager Donna Withers and for each crew member working under her, however, the objective remains to conduct bird counts with what personnel are available and with some measure of consistency: once a week, on the same day, at roughly the same time throughout the pelican's nesting season from February through September.

The purpose of these annual bird counts is to monitor long-term population trends of Anaho's nesting colony. Consistency in counting is necessary to compare Anaho's bird populations with other American white pelican colonies in other parts of the western United States and Canada, because in every one of these locations suitable nesting grounds are disappearing. Broad-range population assessments will allow refuge managers from different regions to see how white pelicans are faring.

"Ten thousand pelican chicks hatched on the island in 2000," Donna chimes in, understandably proud of the high numbers under her management. "In the late 1980s, only five chicks hatched and fledged successfully. We have some ideas about why the numbers receded so drastically, but we can't say for sure."

I already know what Donna can't say. A massive die-off of fish and birds occurred in 1986–1987. The deaths occurred at Stillwater National Wildlife Refuge, a wetland sixty miles east of Anaho Island managed by the same U.S. Fish and Wildlife Service branch office. The pelicans from Anaho Island make round-trip excursions of more than 100 miles to feed in the Stillwater wetland, which sits adjacent to large-scale agricultural fields. The U.S. Fish and Wildlife Service, in an attempt to keep their relationship with local farmers on good terms, never say outright that agricultural chemicals caused the die-off. Stillwater marsh needs water and, according to Donna, the agency is authorized to purchase water rights only from willing sellers. If the farmers won't sell their water or if the price is too high, Stillwater goes thirsty for another year, and there's nowhere for waterfowl and marsh animals to feed and reproduce except in polluted irrigation tailwater. In the 1980s the pelicans ate fish and other organisms that had been swimming in this chemical cocktail. The accumulation of organochlorine pesticides and PCBs (polychlorinated biphenyls) in the birds' blood and tissues weakened their health and caused eggshell thinning.

Bob's bark to "hang on" jolts me to attention. He sets the throttle forward, and the boat picks up speed. Donna and I clasp the side of the boat, and it's a good thing, too. The wind is so strong it punches a lens out of Styron's eyeglasses as he stands beside Bob at the wheel.

"This lake is full of surprises," he shouts, groping for the plastic lens. "Good thing I only need one eye for the spotting scope." Taking the hint, I pocket my specs.

The wind across my open mouth whistles like breath blown against the mouth of a beer bottle. Behind us, Sutcliffe Marina recedes into big sagebrush and rabbitbrush and the bow faces the open water and Anaho's sandy western beach.

"We'll just count the pelicans and get off the island," Styron shouts into the wind. "I don't want to spend another night out there."

The boat engine revs over small whitecaps, slapping a wave so hard that Donna and I, suspended from the fiberglass bench for a mere second, nevertheless come down hard, our tailbones slamming against the hard seat. It feels as if the boat flooring has fallen out, the way the platform of the Gravitron amusement park ride unexpectedly drops away. Styron grabs my shirtsleeve to keep from stumbling. "Better sit down," I say.

As we approach Anaho, an island that emerged sometime within the last 3000 years, I scan the eastern shoreline of Pyramid Lake, a remnant of the late-Pleistocene Lake Lahontan. More like an inland sea, Lake Lahontan covered as much as 8000 square miles during its zenith. When the lake receded as a result of climate change—from damp and cool to arid—it left behind a series of small lakes. These vestiges swell and recede with human activities such as irrigation, which has redirected the flow of rivers and revealed the fossilized remains of camels and horses.

In the 1980s, paleontologists from the Nevada State Museum excavated two camel skeletons and an intact fossilized horse from the edges of Pyramid Lake. These animals may have strayed into deep mud while grazing and got lodged there, unable to extricate themselves and dying where they stood. Mud covered the corpses before scavengers could dismember them.

Bob slows the boat as we near the western shore, the side of the island where the pelicans and other colonial nesting birds don't congregate. No one knows why the birds prefer to nest on the eastern side of the island. Researchers have noted the contrast between the thicker vegetation on the western side of the island and the annually trampled, guano-covered eastern slopes, one of the island's most striking features.

Overhead, pelicans circle the island in an activity called "kettling," spiraling skyward on thermals, swells of warm, rising air that lift the birds to higher altitudes, allowing them to travel to places like Stillwater marsh with less effort. When the nesting season ends in the fall, the pelicans will migrate to the freshwater reservoirs of Sonora and Sinaloa, their wintering grounds.

A welcome wagon of invasive Russian thistle balls blows into my legs, tugging attention from the pelicans as the barbs pierce my socks and scratch thin ankle skin. As I hunch over to yank the barbs from my socks, I pause. I'm stooped over on a stretch of shoreline that stood twenty-six feet underwater in 1913. Dingy rings around Pyramid Lake's shore reveal that the lake's level hasn't always been this low. Paiute fishermen also etched lines onto the face of the pyramid-shaped rock just north of the island to mark water levels between the years 1924 and 1928, when fish populations of cui-ui and Pyramid Lake cutthroat trout declined. The Paiute traditionally depended on these fish for protein, and they were the axis on which their cultural traditions revolved.

The lake's drop is credited to the U.S. Reclamation Service (later the Bureau of Reclamation), under President Theodore Roosevelt. The service constructed Derby Dam to divert Truckee River water from Pyramid Lake to Fallon, Nevada, further east, where white settlers engaged in budding agricultural projects, transforming the West's arid lands into a thriving hub of agricultural activity. The project stretched the Truckee waterway all the way to Fallon, leaving little water to travel its original course.

White lines, similar to those around Pyramid Lake, also encircle the island. Called "whiting," this chalky substance composed of salts looks etched into the island like a tattoo, marking the eighty-foot drop in Pyramid's water level during the twentieth century. When the government surveyor, John C. Frémont, ascended a nearby mountain peak in 1844 and glimpsed for the first time what he named "Pyramid Lake," Anaho Island exhibited only 228 land acres. By the end of the twentieth century, the water level had plummeted so drastically that the island now measures some 750 acres.

As Styron and Donna hand me bird-counting gear—spotting scopes, binoculars, charts, cameras, and the pelican-colony map—I notice whitish gray rocks that look like dirty clumps of baking soda. These boulders, composed of tufa generated from beneath the lake's surface, were exposed as recently as the twentieth century. The lake water, rich in carbonates, combines with the high calcium stores in underground springs around Pyramid Lake, and the chemical reaction produces tufa.

There is nothing common about Anaho Island. It is covered with porous, sometimes chalky rock forms shaped in elegant turrets and globular, convex boulders called "popcorn" tufa. Popcorn Beach, a stretch of the western lakeshore, is known for the tufa boulders chucked helter-skelter as if by Goliath shot-put competitors.

We stash our safety vests in a heap behind a native shadscale shrub, one of the saltiest dogs of Anaho Island, which sprouts from soil so alkaline you could pat blocks of it together and sell them for salt licks. The bush's older twigs have become pointed and sharp, scratching the backs of my hands.

Donna spots the folded skeleton of a pelican chick. "From last year's brood, I think. Probably taken by a gull. Gulls prey on pelican eggs and newborns." I brush my fingers along the thin leg bones hewn by sand and wind. Their surface is smooth. "Pelicans don't tolerate disturbance," Donna continues. "If people get too close to the shoreline, the birds may fly off and leave their eggs and chicks exposed." Pelicans don't abandon their nests permanently, she explains, but just long enough for gulls to pirate nests and the sun to kill chicks with heat prostration.

"Let's snap to it. No time to dither with weather coming in," Bob shouts from the boat. Packs and gear mounted on our shoulders, we ascend the island's greening slopes toward the summit, approximately 550 feet above the lake surface. It has rained recently, and boulders strewn willy-nilly are crusted over with burnt-orange and lime-colored lichens. The swatches of color seem almost garish against the gray rock. We pick a route around the boulders to avoid the classic wheel-and-spokes snare of an orb-weaving spider. These arachnids construct their webs between large boulders and weave silky anchors on either side of the wheel to stretch it lean and tight. We have a difficult time seeing the transparent threads, and the webs snag on our shoulders and elbows, which tear the ornate patterns.

Donna says, "I don't think I've ever seen it looking so green out here. The rain did the island some good. Made it perk up."

Cheatgrass spreading across Anaho Island

"The rain just makes it look prettier if you don't know what you're looking at," Styron says to me. "This is all cheatgrass and red brome," he smirks.

The soft plumes I'd been plucking from my socks were cheatgrass
heads, an invasive species that has not only covered the island but become
widespread throughout the Great Basin. I had seen cheatgrass only in the
pinnacle of summer, when its champagne-colored stalks threatened to
ignite a fire. Today its hypnotic apple-green ripples are soft to the touch
and nothing like I remember.

Range scientist Robin Tausch, who heads several projects, including an integrated approach to weed management and restoration on western rangeland, told me that cheatgrass and its coinvasive red brome dominate the island. Red brome mostly congregates on the southern side of the island, whereas cheatgrass prefers the upper reaches of the island's northern end.

Called "cheat" because it steals moisture and soil nutrients from native plants, cheatgrass was an accidental introduction. It traveled to the eastern United States in a grain shipment from Europe in the 1800s. The plant first arrived in the western states by entering British Columbia and northern Washington as a contaminant in grain seeds and animal feed. Cheatgrass traveled across the West following herds of cattle and sheep. Wherever livestock grazed the nutritious native bunchgrasses to nubbins, cheatgrass found a niche and established.

But if there have never been cows on Anaho Island and only a few goats grazed for a brief time decades ago, how did cheatgrass get here? Cheatgrass seeds may have stowed away on the feathers of nesting pelicans when they fished infested wetland areas of the mainland and then returned to the island to feed their chicks. Wind and storms may have carried them, or the seeds could have hitchhiked to the island's shores by clinging to the clothes of boaters who illegally landed on the island. They also could have ridden the pant legs or shoelaces of a well-meaning government refuge manager, or embedded themselves in the ankle socks of a volunteer like me.

Rust-colored red brome drapes the landscape in a pinkish hue. Red brome, native to Mediterranean Europe, was also an accidental introduction in the mid-1800s, arriving as a contaminant in grain shipments. Like cheatgrass, red brome has spread to many areas of the western United States, replacing native vegetation, altering ecological processes, and negatively affecting wildlife. For example, the spiky florets of red brome sometimes become embedded in the eyes of the red-shouldered hawk as it hunts small mammals and reptiles that inhabit vegetation. With its vision impaired, the hawk cannot hunt effectively and may die of starvation.

Although they are annuals, cheatgrass and red brome sometimes seem immortal because several generations grow, spread their seed, and die within a single season. Tausch made it clear that not only do these two invasive species dominate the vegetation on Anaho, but except for spotty growth of a few other native plants like shadscale, very little native vegetation remains on the island. In fact, there are only two individual sagebrush plants left.

Invasion ecologists have discerned that the majority of alien arrivals cannot compete with indigenous inhabitants for resources. They either die out or remain confined to a circumscribed area. Alien plants that do establish hardy populations, the way cheatgrass and red brome have done, have cleared all the hurdles, including extreme arid conditions, trampling or overgrazing by animals, and competition with native plants for space, nutrients, water, and sunlight.

Environmental resource scientist Bob Nowak told me, "I don't think we'll ever completely eradicate cheatgrass from North America. It's too widespread and abundant to ever hope for that. We may be able to minimize its hold in certain areas, but it's going to be very expensive."

Kim Allcock, postdoctoral fellow with the Department of Natural Resources and Environmental Science at the University of Nevada, Reno, was even more adamant. "Eradication is not an option with cheatgrass—it's too widespread," she said. In Nevada alone, this invasive grass covers 25 million acres and continues to spread. Allcock is currently involved in several experiments, one of which tests several native plant species along with a few exotic ones such as crested wheatgrass, Siberian wheatgrass, and a sterile wheat/rye combination to see how well they are able to reduce cheatgrass dominance. "It's very difficult to move from a cheatgrass system to a native community," she explains, so the idea is to create an interim step. Eventually, range managers would replace the experimental plants with reintroduced native plant species, giving them a jump start on cheatgrass.

Another of Allcock's experiments tests ways of shifting the balance of competition between native plants and cheatgrass by altering soil nutrients. Researchers apply labile carbon to the soil, which encourages microbes to take up nitrogen, leaving less of it for cheatgrass growth. Simultaneously, Allcock and her research team are planting a variety of efficient native species that maximally use available resources. Their findings show that reducing nitrogen has dramatically decreased cheatgrass seed output and biomass.

Because cheatgrass can't be entirely eliminated, Allcock explained that the goal of these experiments is to create a native plant community in which cheatgrass is a smaller component rather than a dominant feature of the landscape. "It is a matter of shifting ecological processes and the balance of competition in favor of the native species, and making cheatgrass a better-behaved and less dominant member of the community."

Anaho Island may be an ideal location for testing different native-species restoration methods. Robin Tausch agreed that the western portion of the island, the side of the island nesting birds don't use, could be a good site for such experiments. Anaho is a simple system. It is small and therefore has fewer species than larger islands. There are deer mice, lizards, seasonal nesting birds, rattlesnakes, and what remains of native vegetation. Researchers would get rapid feedback about whether their experiments are working, and the risk of reinvasion is less than on the mainland. So why hasn't the island been considered for such experiments? Tausch answered with no hint of humor, "It would require interagency cooperation and money, both difficult to come by."

Styron and I pause to shed layers of clothing. Although it's overcast with clouds rolling in from the west, the morning is heating up. I pencil notes into my pocket book. Styron eyes me playfully, "I keep forgetting you're not a scientist. You're one of those literary types—always writing down what comes out of people's mouths. I guess I better watch what I say."

We round a bend, our boots crunching on chips and splinters of tufa. When we reach the overlook below the island's summit, we drop our packs on the soft stems of cheatgrass and eat cheese sandwiches in silence. Perched about 200 feet above Anaho's pelican colony along the eastern side of the island, we've risked bad weather and rattle-snake bites for this count, so we finish our lunches quickly and get on with it.

We set up our spotting scopes and divvy up the subcolonies that make up Anaho's pelican colony. It's like a city of well-defined neighborhoods. Donna pairs me with Styron so I can learn counting and recording procedures, and the two of us take subcolonies A, B, C, and E. Ridge subcolony on the spit of the island's southeastern side is empty, and so is Bluff, immediately below this overlook. It's just a bare area of whitish dirt where the pelicans nested in previous years. "Fewer pelicans this year," Donna says. "There's just not enough to eat because of last year's drought. Hopefully this year will be a wetter season."

These pelican neighborhoods remain the same most years, and no one really knows how the birds choose one over another. Each area is used intensely, and when the birds' guano dries to a white powdery substance, it becomes acidic and burns any vegetation attempting to re-establish there. Even in the off-season, these areas rarely revegetate. When the birds return, they simply take up residence in the old subcolonies, highly visible by their nakedness.

Styron presses his good eye against the Prominar lens, panning the spotting scope to get an overall picture of the colony. The brim of his white canvas hat flat against his forehead, the chinstrap cinched tight against the wind, he says, "Yeah, thin year." With the metal hand counters set at zero, Styron and I begin our counts. We count each subcolony independently and then compare numbers. For every ten pelicans, I push the counter's tap. Ten. Click. Ten. Click. Ten. Click.

In the 1950s, bird counters didn't keep the kind of distance from the birds that we do today, and pelican-banding missions didn't operate with the knowledge of pelican biology and behavior that they possess now. Back then, managers had planned to band the chicks in the middle of the day. When they landed on Anaho, the parent birds flew off raving mad, leaving their chicks exposed to desert sun and predation by California gulls.

Management agencies meant well. Banding would allow them to track the pelican's life cycle from chick to adult and develop more effective ways to manage and preserve the population. But the crew didn't know much about pelican parenting behavior or chick physiology and stress tolerance. Banding can't occur during daylight hours because one parent bird always guards the chick from sun and gull predation. The parent straddles the chick, using its body to shade the chick from overexposure. While one pelican shades the chick, the other parent fishes, and then they switch in what is called a "relief ceremony," an elaborate, sometimes drawn-out affair when mates exchange places on the nest, either guarding eggs or shading chicks.

I look up and see seven pelicans stream by at eye level, drifting like a gossamer thread on the wind stream. "Those guys are still prospecting," Styron says, thumb paused on the counter button. "They haven't found a mate or a subcolony to nest in yet." Absorbed by the line of pelicans winging past us, I lose my count on subcolony C and have to start over. Pelicans are distracting birds because they are so elegant in flight. They are at their most graceful in the air and certainly while engaged in synchronized feeding. On land, they are oafs, bumpkins, weeble-wobbles. The pelicans loafing in dense congregates look like doddering old men, preening their feathers and clap-clapping their bills at one another.

Getting sidetracked while counting isn't a big deal, but other people have been distracted while watching pelicans, and bad things have happened. During National Safe Boating Week some years ago, two pleasure boats were motoring together toward the western end of Pyramid Lake, when the first boat stopped to let out a water skier. The second boat's driver, entranced by the antics of a group of pelicans alongshore, didn't see the first boat slow, and his larger craft rammed into the stern of the smaller vessel. A woman onboard the first boat incurred blunt trauma from the collision and died.

I begin the recount of subcolony C, numbering pelicans as they rest with their bellies pressed to the cool sand, while others hem and haw and stand on their eggs, the shells now strong enough to withstand the stress. Occasionally, the birds flutter their gular pouches, the stretchy skin of the throat, to cool down. There are other birds strutting around in pairs or small posses, something G. B. Schaller described in 1964 as the "strutting walk." The birds look like swaggering desperados from *The Good, the Bad, and the Ugly*, an outlaw band of birds advancing across the dusty landscape, their heads held rigid, bills pointed downward, stern and full of business.

Pelicans don't make many vocalizations. The only sounds I hear are

American white pelicans fly over Anaho Island during the breeding season.

grunts from their nests, a typical noise they make when tending eggs and chicks. It sounds something like a cow mooing. If I crouched closer to them, their snapping bills might sound like the clink of castanets, or maybe they would have a throatier sound, like the wallop of a thunderclap or the bottoms of a pair of shoes clonked together. More than vocalizations, pelicans use their wings and bills to express themselves. To display aggression, a pelican often jabs its bill in the direction of the offender and snaps it closed several times. Pelicans also flap their wings in a simple stretch to show displeasure or anxiety.

We complete the bird count and fill in the numbers for each category on the tally sheet: numbers of pelican adults on each nest, numbers not nesting, numbers prospecting for mates, numbers of eggs and chicks visible, numbers of pelican chicks unattended by an adult. Donna also adds up cormorants, and Styron notes the number of great blue herons, totaling the number of nest clusters scattered here and there in patches of native greasewood. Then we gather our gear, shove it into our packs, and file off the summit.

The three of us are silent as we climb down a steeper section where the trail meanders between two lenticular tufa crags, named for the way the tufa stacks in chockablock fashion, some rounded hunks bulging out in a fan shape, others straight and square as horses' teeth.

As we wade through the cheatgrass, Donna indicates several boulders on the hillside where the green stems of this exotic grass have formed haloes around the rocks to catch shedding moisture. A common strategy for exotic plant species is to grow near water tanks and under the short, dense canopies created by native shrubs in order to filch seeped, dripped, and otherwise unused water. The spikelets at the top of the cheatgrass stems droop, nodding like the heads of tired travelers, but this grass is anything but weary in its foreign home. On Anaho Island, where precipitation averages a mere twelve to eighteen centimeters each year, it's a free-for-all for cheatgrass and red brome, which are adaptable to almost any set of environmental conditions and can absorb even a modicum of moisture from the soil.

Donna assures me that cheatgrass and red brome don't negatively affect the pelicans. In fact, she says, vegetation type is of little importance to them because, although they nest around vegetation, their actual nests are mere scrapes in the sand.

But I wonder if such a narrow conservation focus on the white pelicans might not ultimately come at the expense of the very bird the agency hopes to preserve. Summer is fire season in the Great Basin, when cheatgrass stalks blaze like matchsticks. Before the arrival of these introduced grasses, native shrubs and bunchgrasses in the Great Basin were spaced well apart, and fires occurred about every fifty years. When cheatgrass and red brome established, their dense stands of erect stems filled the nooks and crannies between native shrubs, providing abundant fuel for fire. As Allcock explained, "When fires occur in stands of native bunchgrasses and shrubs, they don't provide a continuous cover of fuel to carry a fire for long distances. With cheatgrass present, there is a continuous carpet of fine fuels." During the summer months, the crispy stems of red brome and blonde cheatgrass burn like paper and carry flames from one native plant to another. Now fires careen through the region every ten years or less.

These fires are high-intensity fires, mercurial and dangerous to the people fighting them, as well as a crippling expense for the federal and local governments. In August 1999, dry lightning struck several areas in the Great Basin, flash-panning 1.7 million acres of high desert, including homes, at an estimated cost of $13 million. Aside from the expense and the hazards of putting these fires out, the damage is terribly difficult to repair.

Let's say there's a lightning strike in the knee-high summer cheatgrass and fire sweeps through a desert area, burning everything in its path. After a fire, soils are temporarily high in nutrients, and the rapid reproduction rate of cheatgrass and red brome allows them to suck up available nutrients before the slower-growing native plants have a chance to establish. In other words, fire helps cheatgrass expand its range.

According to Robin Tausch, "It's not a question of *if* but *when* a fire will happen on Anaho Island." University of California, Davis, wildlife biologist Dan Anderson maintains that the pelicans would be unable to nest during the year in which a fire occurred on the island. If a blaze occurred when their nests were already established, a fire would ravage the nesting colony and obliterate the season's brood. And what about the island's native plants? What about its year-round animal residents?

I lead the group along the Narrows, an intermittent trail strewn with boulders, where I nearly step on a sunning rattlesnake coiled in a shallow depression on top of a boulder. It blends well with the spongy grayish lichens, and to me, it looks no different from them. I stop suddenly, and Styron pitches into me. The snake doesn't wake in the commotion. The length of its body below the head remains coiled in a backward "S" shape. Olive-colored smudges spaced evenly along the top of its body look like pudgy arrows pointing in our direction. A bulge protrudes from the snake's midsection, possibly a native deer mouse, *Peromyscus maniculatus*, a name meaning "mouse with little hands," the only mammal on the island. Like all snakes, this rattler scented the mouse with its forked tongue, tracking odors along the ground. Unable to chew because its teeth are like bowed needles, the snake swallowed the mouse whole by unhinging its lower jaw.

We three gawk at the snake, awed by its flawless skin patterns and almost Zenlike languor amid the din of our heels pounding the dirt. Although they may react with enthusiasm and pleasure to pelicans, most people are less likely to get distracted by the beauty of a rattlesnake than they are by their fear of a bite from one. In fact, most people are more likely to attempt to kill a rattlesnake than to leave it alone.

Specialists like vertebrate ecologist Kyle Ashton do get excited about snakes, especially the snakes on Anaho Island. I contacted Ashton after my first experience on Anaho and asked him about the rattlesnake there. He eagerly explained that he wants to do more genetic work with Anaho's rattlesnakes, but that research funding has run dry. He suspects the island population may be a unique subspecies of the western rattlesnake or even a species in its own right—an island endemic.

Cheatgrass and red brome enter stage right. Tausch told me that, although a lightning strike to the island is unlikely because it is small and surrounded by water, "I'm concerned about recreational boaters stranded on the island. When some boater's engine fails and they get stuck on the island and light a cigarette or start a campfire, the whole island will go up, and the last of the island's native vegetation will be destroyed. No natives will come back after a fire there. Cheatgrass and red brome will beat them out." A self-perpetuating cycle follows: both invasive grasses grow back

Anaho Island rattlesnake, a possible island endemic threatened by invaders

in greater abundance following a fire, and more cheatgrass and red brome lead to more fires.

Ashton told me that the long life span and slow rate of reproduction in rattlesnake species make them particularly vulnerable to environmental disasters. For example, take the story of a rattlesnake population ravaged by a wildfire in Arizona's McDowell Mountains. I spoke with Steve Beaupre, assistant professor of biological sciences at the University of Arkansas, who studies a population of the western diamondback rattlesnake native to the desert near Scottsdale, Arizona. In the summer of 1995 he and his colleagues Dave Duvall and Jack O'Leile witnessed a severe wildfire that swept through their research sites. The fire started when lightning struck in the vicinity of the McDowell ridgeline. Red brome, cheatgrass's coinvasive on Anaho Island, fueled and carried the fire.

Beaupre reported that more than 20,000 acres burned. The snakes were hidden in burrows and pack-rat middens, which Beaupre described as "incipient bonfires." The middens smoldered. He depicted the research sites as moonscapes following the fire and the middens as mere "pits of ash" that acted as rattlesnake traps.

The McDowell blaze was devastating. If red brome hadn't been a significant presence in the region, providing excess fuel and destroying native fire-intolerant shrubs, the fire would not have covered such a large area nor spread so rapidly. While the western diamondback rattlesnake is common in Sonoran desert scrub and was "only" locally exterminated in what might be thought of as a 20,000-acre island habitat, a fire on Anaho, about 750 acres in size, would severely reduce or possibly wipe out the island's population of rattlesnakes, deer mice, and lizards.

Whether or not the rattlesnakes on Anaho are endemic, we'll never know their place in the island's ecosystem unless they are studied by people like Kyle Ashton and Steve Beaupre. If the experiments to restore health to the Great Basin's native vegetation could take place on the island, perhaps they'd be successful and a fire could be avoided. With unprecedented concentrations of invasive plants on Anaho Island and the surrounding mainland, managers have sufficient reason to reconsider their management and research focus. In fact, the island meets the Bureau of Land Management's criteria for restoration: an unburned area highly vulnerable to invasive species and possessing "high potential to attain proper functioning condition."

I glimpse four individuals of fourwing saltbush, their roots cleaving to a plot of soil despite stalks of cheatgrass crowding them. A single stalk of native Nuttall's larkspur catches my eye. The plum-colored flowers remind me of jack-in-the-boxes with funnel-shaped spurs like tapered purple stocking caps. Nearby there are still a few squat, grayish white winterfat shrubs, often called "moldy wool."

Donna and I look skyward. Through binoculars slung around my neck, I see hundreds of pelicans corkscrew upward as if in a wind tunnel. The shape formed by these soaring birds has the character of a helix. The kettling pelicans soothe my mind as they twine, the lowest birds vanishing behind the summit as they pass, around and around. Then they turn, exposing only their white backsides, and disappear from my sight, those stark white backs blending into a background of cumulus clouds. On the next rotation, they expose their undersides with the telltale black primaries and reappear on my radar screen. I can also distinguish the primary feathers on the undersides of the pelicans' wings. They extend outward to five separated feather tips, distinct as fingers. The pelicans' retracted S-shaped necks exaggerate the square horny knobs on the tops of their bills. These elegant birds remind me of something antediluvian, something prehistoric. In fact, there are pelican fossils dating back 40 million years.

Donna and I catch up with Styron and finish the descent to the boat. The wind calms, so Donna suggests we circle the island to get a different view of the pelican colony before heading back to the marina. "I might not get this view for a while," she says. Bob steers the boat clockwise around Anaho. When we reach the northern tip, I can make out Pelican Point and Wino Beach on the mainland to the north, where whistling swans congregate in late fall and early winter. Blonde with cheatgrass, the treeless desert hills of the island are blinding in the unleashed sunlight, a stark contrast to the aquamarine of Pyramid Lake.

After making a loop of the island, Bob steers the boat toward the marina. As we cross the lake, I watch the curl of whitewater waves crest against the bow of the boat. Two double-crested cormorants fly low overhead.

"It's a nice lake, but it ain't the ocean," Bob shouts into the wind.

"He didn't really mean it," I whisper to the Water Babies under my breath. "Don't pay attention to him. He's cranky." I pay homage to those bairns of the lake—just in case.

Chapter 2 Pister's Pupfish

We knew the Owens pupfish would go extinct that day if we didn't act.
—Phil Pister

On a searing August day in 1969, just one month after men landed on the moon and returned home by cannon-balling into the ocean, fisheries biologist Phil Pister worked for hours, even into the night, on his own time and without pay, carrying buckets of tiny fish from one pool to another in the California desert. Usually we speak of humans driving a species to extinction. In Phil's case, he drove them, in his pickup truck, to their salvation.

Ash Meadows National Wildlife Refuge, located seventy-odd miles northwest of the chink and jangle of Las Vegas and adjacent to eastern California's Death Valley, exists today because of work Phil Pister and other biologists did during a time when few cared about the fate of a non-game fish. In Ash Meadows, thirty spring-fed pools and marshlands burble with warm subterranean waters. These thermal baths function as islands of water surrounded by a sea of steppe—arid, shrubby expanses of land. Biologists call these inside-out islands "reverse islands" because, like traditional islands, bodies of land surrounded by water, aquatic islands are equally isolated, but hemmed in by endless stretches of land. The warm pools exist as one of the most remote and least understood habitats on Earth. These islands teem with plants and invertebrates and provide a home for rare creatures known as pupfish.

I realized that rather than detract from the original exploration of islands and invasive species, the pools will break open the common definition of an island and expand it. These strange islands and their finned inhabitants, called "pupfish" for their playful behavior and diminutive size, have attracted biologists' attention for nearly a century, prompting them to defend these odd fish against exotic predators despite a lack of money or institutional support. To better understand how these aquatic islands and their residents rouse resource managers to their unflagging defense, I contacted Phil Pister. During his thirty-eight-year tenure with the California

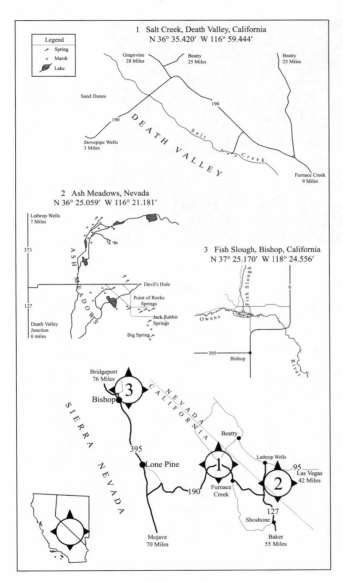

1 Salt Creek, Death Valley, California
N 36° 35.420′ W 116° 59.444′

Legend
Spring
Marsh
Lake

Grapevine
28 Miles

Beatty
25 Miles

Beatty
25 Miles

Sand Dunes

190

190

DEATH VALLEY

Salt

Creek

Stovepipe Wells
3 Miles

Furnace Creek
9 Miles

2 Ash Meadows, Nevada
N 36° 25.059′ W 116° 21.181′

Lathrop Wells
7 Miles

373

ASH MEADOWS

127

Death Valley
Junction
6 miles

Devil's Hole

Point of Rocks
Springs

Jack Rabbit
Springs

Big Spring

3 Fish Slough, Bishop, California
N 37° 25.170′ W 118° 24.556′

Fish Slough

6

Owens

395

Bishop

River

Bridgeport
76 Miles

3

Bishop

NEVADA
CALIFORNIA

Beatty

Lathrop Wells

95
Las Vegas
42 Miles

SIERRA NEVADA

395

Lone Pine

1

2

190

Furnace
Creek

127

Shoshone

Mojave
70 Miles

Baker
55 Miles

Salt Creek in
Death Valley, Ash
Meadows National
Wildlife Refuge,
and Fish Slough.
Illustration by Rick
Moser

Owens pupfish. Photograph by Phil Pister

Department of Fish and Game, Phil has stewarded nearly 1000 bodies of water in which many of the sixteen species of pupfish native to the American Southwest make their home. I visited two major sites of Phil's conservation efforts, Fish Slough and Ash Meadows National Wildlife Refuge.

Phil and I met at his home in Bishop, California, to retrace the steps he took thirty-five years ago to save a dying species. We had spoken so many times on the phone about his work and life that when we finally came face to face on that early Sunday morning, we shook hands and hugged as if we'd known each other many years. Clasping his hand, I felt the deeply etched lines of his palm, fingers thick with years of rugged outdoor work. Phil was dressed in jeans and a sweatshirt picturing the ancient, gnarled bristlecone pine, custodian of the White Mountains to the east of us. His bushy gray eyebrows lifted above warm brown eyes as he smiled, and the creases in his cheeks deepened, a testimony of his seventy-seven years and of many seasons working in the desert sun. He wore a blue baseball cap with the logo "UC White Mountain Research Station" to protect his head, where only a whisper of hair remained. I gripped the hand of a man who had spent nearly half a century tending to the health of a habitat.

To prepare for my meeting with Phil, I had decided to visit another site of his conservation efforts farther south at Ash Meadows refuge. My colleague and friend, Crystal Atamian from the University of Nevada, Reno, accompanied me down Highway 95 toward the refuge and another set of reverse islands.

Hot gusts blasted through the open windows as Crystal and I passed through the mining town of Goldfield, where the flat timber dressing on homes and shops splintered in the desert sun. The landscape of western Nevada appeared stretched like the tanned hide of a deer, its surface a uniform russet devoid of any shade trees or even a jumble of boulders. We freewheeled through Scotty's Junction and Springdale, keeping the Bullfrog Hills to the west and the vast sweep of the Nevada Test Site to the east. Dry air boxed my ears as we passed Angel's Ladies Ranch with its hodgepodge of trailers. Painted across the front of one champagne-pink mobile home, "Brothel" appeared with the same bald frankness of a "no parking" sign.

Aside from hookers, slot machines, and hot weather, Nevada's least celebrated but perhaps most important asset remains its singular plants and animals. In Ash Meadows, biologists such as Phil Pister have identified at least twenty-four endemic species of plants and animals—species that live nowhere else but on this refuge. Pupfish, some the size of a pen cap, account for three of these unique species.

The depth of the pupfishes' watery island homes remains a mystery. Divers from the U.S. Geological Survey, while attempting to measure a cavernous spring called Devil's Hole, descended 300 feet before abandoning their mission. They never touched bottom. Later, two civilians lost their lives during an illegal night dive when, disoriented in the cave shafts, they couldn't find their way back to the mouth. Without safety cables to haul them out, the men ran out of oxygen. Their bodies have never surfaced.

This tragedy, as well as a dearth of funding, has impeded exploration of these reverse islands. In the last decade, we may have learned more about the forbidding Martian atmosphere than we've learned in the last century about the ecology of these peculiar islands. The concept of endemism, the evolution of plant and animal species unrepeated anywhere else in the world, has its clearest meaning here. The pupfish of this desert archipelago, like the Galápagos finches studied by Charles Darwin, are "indigenous productions." They developed into uncommon species as a result of geographical isolation, spawning in the pools when the dire wolf and saber-tooth tiger stalked camels here.

Despite their highly specialized existence, these tiny fish outlived their fierce mammal neighbors and endured the sweeping environmental changes of the last ice age, when what is now southern Nevada transformed from a cool, wet climate to the present hot and arid one. Because evolution on islands occurs faster than on mainland or non-island habitats, island biogeographers can study how evolution works through the remarkable adaptations of species like pupfish, and then apply these principles to the slower evolutionary changes of non-island species.

So why, after surviving dramatic climate changes and outliving the seemingly more robust saber-tooth tiger, have these fish declined to the brink of extinction in the last 150 years? Phil began to address that question by telling me that people often ask him why he's so concerned with extinction. "Extinction is a natural process, so what's the big deal? I tell people that I split extinction into natural versus human-induced species loss." Phil also tells them that human disturbances like the introduction of nonnative species or the siphoning of groundwater for housing developments and agriculture drive species like pupfish to a premature end.

By noon, Crystal and I slowed at Lathrop Wells and forked onto Death Valley Junction Road to within chucking distance of the state line, then turned our backs to California and headed east on Bill Copeland Highway, the gravel road through the refuge that shears off its southwestern wedge. This belt of the Mojave Desert surrounding Ash Meadows possesses characteristics of both the Great Basin and the Sonoran Desert. Big sagebrush, creosote bush, and fourwing saltbush would appear familiar to anyone living in northern Nevada, while compass barrel cactus and claret-cup cactus might seem like home to those living near Phoenix.

The two-wheel-drive pickup jounced and fishtailed as the tires spit rocks into creosote shrubs edging the road. In the rear-view mirror I spotted a whorl of dust, plant stems, and dried leaves billowing behind us. We parked in the empty parking lot of the U.S. Fish and Wildlife Service's Ash Meadows National Wildlife Refuge. Crystal walked up the wooden ramp leading to the entrance of a double-wide trailer and knocked on the door. No one answered, although it was two o'clock on a Wednesday afternoon, the middle of a regular business day. Crystal shrugged, "Nobody home?" and rapped a second time.

A breeze brushed my feet as I craned my neck to skim the view over my shoulder, a wind-rattled port-a-potty tucked at one end of the trailer and broad meadows of crystallized salt and chalky knolls of borate rolling out of sight in all directions. "The gold rush wasn't the only mining hustle in Nevada and California," I said. In the 1800s, men using twenty-mule-team wagons quarried three tons of borate minerals each day. We continue to use borates as a detergent booster in laundry soap and as a primary ingredient in agricultural insecticides. Manufacturing fiberglass insulation also requires the use of borates, and some older public restrooms still dispense Boraxo, the grainy white soap powder issued from a pump.

Piles of corkscrew-shaped mesquite fruits littered the wooden planks under my feet, as Crystal and I noticed a handwritten message on the windowsill. Embossed with a U.S. Fish and Wildlife Service logo, the memo apologized for the visitor center's closure and cited budget cuts restricting full-time staffing. The sign also stated that any questions about the refuge could be e-mailed to staff, who would answer as promptly as possible.

"Welcome to the cyber visitor center of the twenty-first century," Crystal said, as she plodded down the ramp to an information kiosk next to the station.

I scooped a handful of mesquite pods and followed her. Wind whisked dust and the musty scent of creosote's resinous blossoms across the parking lot. From my studies of desert flora, I recognized groves of velvet ash, the namesake of the refuge, and screwbean mesquite that dotted the area around the visitor center. I noticed gumweed, a plant listed as threatened in refuge literature. "Hey, check out that butterfly on the gumweed," Crystal said, brushing strands of windblown hair from her lips. She pointed to a small butterfly hovering over the plant's sticky yellow flowerheads. "They may not have any staff, but at least the refuge doesn't let cattle in," she said. "The wildflowers are safe."

These "flutter-bys," as I'd grown up calling them, are salt-grass skippers, named for the caterpillar's host plant salt grass, which has the texture of straw, each blade stiff as a picket. When the skipper alighted on a yellow-green gumweed stalk, it rested just long enough to let us view the color of its wings, a mixture of lemon peel and dark orange, each veined wing framed by a thin black line. Suddenly, ten or more of the same type of butterfly appeared in loping flight between flowers. As I watched the diurnal insects, I recalled "The Butterfly," a poem by Vermont poet Louise Glück:

Look, a butterfly. Did you make a wish?
You don't wish on butterflies.

You do so. Did you make one?

Yes.

It doesn't count.

Eager to glimpse the pupfish, Crystal and I wandered toward King's Pool, the first restoration project in the refuge's history. As a result of labor-intensive restoration efforts, King's Pool hosts 90 percent native fish compared to its original 90 percent invasive fish. We followed the smell of mown salt grass and warm mineral water wafting toward us on the pathway. Freshets babbled beneath a grove of native ash trees, their numbers dwarfed by large clumps of invasive tamarisk trees that steal valuable water from the springs. Native prince's plume grew skyward out of the salty soils along the footpath. Its tall, wandlike stems ended in a torch-shaped flowerhead the color of butter. Ants climbed these columns like mountaineers and topped out on a silky ledge gleaming with sugar.

Sedges, with their smooth, three-sided staffs, formed a hedgerow around the spring. Their shoots chafed and hummed the sound of rosined horsehair against the D-string of a violin. We listened, and then Crystal recited the rhyme used to help students of wetlands ecology distinguish between the often lumped-together "swamp grasses."

"Sedges have edges. Rushes are round," she chanted.

"And grasses are hollow right down to the ground," I chipped in.

We gazed into the cerulean water, marked with an "off limits" sign. The refuge prohibits swimming, fishing, snorkeling, diving, and even dipping toes into the warm pools. On that blistering day, it felt like peering over the backyard fence at the neighbor's pool party you haven't been invited to.

But these meadows haven't always been off limits to people. Prior to June 1984, when the federal government designated Ash Meadows a refuge for endemic fish, insects, snails, and plants, human communities made these grasslands their home. Evidence of human occupation in the form of house circles, rock art, storage pits, and burial sites tell archaeologists about the humans who occupied the steppe country as early as 2600 B.C.

"Except for nuclear waste," Crystal said, "I can't imagine modern humans creating anything of interest that would last 4000 years or more."

I squatted at the rim of the limestone pool and pressed my palms into mortar depressions made centuries earlier when Desert Shoshone ground corn to flour. A nomadic people who moved into the area about 1000 years ago, the Shoshone must have ground corn kernels as streaks of blue iridescence wiggled in and out of their view. The Shoshone congregated here to gather mesquite fruits and piñon nuts and to hunt rabbits and deer stirring meadow grasses. The people would have witnessed pupfish courting behavior as Crystal and I did. Noiseless at the edge of the spring, which was blue as the kiddy pool I had as a child, I saw the glimmer of a male Amargosa pupfish, one of three endangered species inhabiting the refuge.

I trained binoculars on the Amargosa, his shimmering scales the color of early blueberries, each fin outlined in charcoal as if with a fine paintbrush. Then the breeze picked up and ruffled the surface water, erasing my

A spring-fed pool in Ash Meadows National Wildlife Refuge inhabited by endangered, endemic pupfish

view like a shaken Etch-a-Sketch. When it died down, I found him grazing along the limestone ledge slick with algae, the pup's staple food. A silver and olive green female aroused him, and he shadowed her movements. Another male approached, equally aroused, and the first male head-butted him. Pupfish are tiny, but tough. The intruder circled, attempting to edge nearer the female, but the increasing ferocity of the first male's charges caused him to retreat. The successful suitor fed near the female, then sidled up to nudge her flanks. This stroking coaxed her to release her eggs. Triggered by the male's advances, she swam to an area of sand and settled there, likely expelling her eggs for the male to fertilize.

Crystal slogged through the sedges to the other side of the pool, and I plunked down in the salt grass. "This water is cleaner than any I've ever seen," she said. The water's translucence belied the true distance to the pool's bottom. It appeared within arm's reach, as if we might rake the sandy bottom with our fingers, but the fish pond's depth likely measured somewhere between twelve and fifteen feet.

Something too large to be a pupfish crept into my field of view. "Hey, that's a crawfish," I said, getting to my feet and pointing at the sky-blue and burnt-orange crawfish gleaming against the sandy bottom.

"Yeah, looks like a lotta them down there."

"Damn. Once you see one, you start seeing them everywhere."

As if the crawfish felt the weight of our stares, they scudded across the sand to hide beneath the travertine shelf. While pupfish have been residents of Ash Meadows for thousands of years, this Louisiana or red swamp crawfish is a new tenant who moved in sometime during the 1960s or 1970s and started terrorizing the neighborhood. It digs like a lonesome backyard dog, burrowing into the springs' walls and weakening the banks, which then collapse more easily. The crawfish is also a voracious eater, severely pruning aquatic plants and eating pupfish eggs and young with abandon. It even consumes adult pupfish when it can catch them, hunting at night when pupfish settle on the pools' sandy bottoms to rest. Due to centuries of isolation, the pupfish haven't evolved any defenses against this

Invasive Louisiana crawfish prey on pupfish.

eating machine. Researchers have noted that this crevice-dweller, when introduced to habitable waters like the warm pools of Ash Meadows, becomes top carnivore in the community and the keystone species, an animal that determines the evolutionary direction of an ecosystem, which may include the elimination of one or more species of pupfish.

How did the crawfish, a freshwater crustacean of Louisiana's swampland, get to the Nevada desert? They certainly didn't crawl all the way from Atchafalaya Bayou. Sharon McKelvey, Ash Meadows refuge manager, suggested that ranchers introduced crawfish for a food source. In Breaux Bridge, Louisiana, *la capitale mondiale de l'ecrevisse* (the crawfish capital of the world), restaurants serve boiled crawfish, or "mudbugs" as they call them there, with cocktail sauce, lemon, butter, and garlic. Crawfish are an international industry, with shipments to France, Spain, and Italy, where they've outcompeted native crawfish species, not only in popularity but also for habitat.

Distributors may have viewed Ash Meadows as a string of free aquariums where crawfish could be raised for the marketplace. But in the western United States, while interest in crawfish cuisine never took off, the crawfish population certainly did. In fact, crawfish numbers have swelled to where a single trap can catch more than 100 of them in only a few hours. Such trapping, although it helps, hardly makes a dent in the population.

Phil had sighed when I mentioned the crawfish. "The pupfish exist despite crawfish," he had said. "We'll never get rid of them unless the government hires a private company to come in and eradicate every last one. But of course, there's no money, so we just have to keep on top of their numbers. At this point, that's all we can do."

As Crystal and I wandered around King's Pool, burnt-umber, yellow, and blue crawfish scuttled along the bottom. Sometimes I saw only their shadows as I knelt at the water's edge, wiggling my fingertips in the spring to see whether one of them detected the movement. Before I had time to reconsider the stunt, a pincer thrust out from beneath the algae-matted shelf and scissored my thumb. I jerked my hand back and the crawfish with it. The creature wouldn't shake loose, so I whacked it against the ground until it relinquished its hold.

It was my own fault. I knew the carnivorous nature of the red swamp crawfish. I stared at Louisiana's state crustacean in the dirt at my feet and imagined it swimming in creamy bisque. I could have eaten it, and there would be one less crawfish to prey on pupfish and lay eggs, which number between 100 and 700 per year for a single female crawfish. Or I could have stepped on it. But I hate directly killing anything, even exotic species, and ended up simply watching it bake to death in the sun.

Worldwide, fresh and saline pools and streams support 100 species of pupfish, most in North America, with some species living along the coast of Venezuela. All sixteen native to the American Southwest remain listed as endangered, threatened, or as species of special concern, with the exception of the Monkey Springs pupfish in Arizona, which is extinct. These pups live in varying conditions, from spalike waters hovering around 100°F to aquatic haunts that plateau around 113–114°F. While some pupfish prefer warm water, other members of the southwestern group exist in streams and rivers flowing at or near 32°F. Some even survive under the ice, while awaiting spring thaw. Although individual pupfish species live at opposite ends of the temperature spectrum, each exhibits an impressive tolerance for wide thermal fluctuations in its habitat, earning pupfish the admiration and bafflement of biologists.

Pupfish present biologists like Phil Pister with a paradox as well as a lifetime of difficult work. Tolerance for water-temperature variation makes pupfish ecological generalists, to an extent, and therefore more robust than many other endemic species. As endemics and habitat specialists, however, pupfish remain vulnerable to human-induced changes in their communities. Drastic changes in water level that expose or completely dry out their food source would cause a massive die-off.

In the early days of wildlife management, protection of sport fishes such as trout—species that people could take home and fry—characterized the work of biologists. But after Phil had been with California Fish and Game for a few years, the vigor of these tiny pupfish and their enormous vulnerability illustrated a larger ecological picture, and Phil began a kind of revolution against conventional fish management devoted to economic gain. A few other enthusiastic and hardworking researchers also sought to change agency policies that favored introduced game fishes like rainbow trout and largemouth bass (exotic fishes Phil calls "chainsaws with fins") over native pupfish.

Phil told me that sport fishermen still introduce largemouth bass, although it's now against the law. The Bureau of Land Management, the U.S. Fish and Wildlife Service, and the Nevada and California Departments of Fish and Game do their best to keep up with these constant reintroductions. They gillnet, scuba dive to spear the bass, and electrofish. (Commonly called "shocking," electrofishing directs a DC current through the water and immobilizes the fish for collection.) This interagency cooperation just manages to keep the numbers down.

In 1948, when Phil was still a wildlife conservation and zoology student

Fish Slough in eastern California is Owens pupfish habitat.

under A. Starker Leopold, biologists pronounced the Owens pupfish, a resident of Owens Valley just north of Death Valley, extinct. But the pathway to extinction was paved much earlier than that. In 1902, following the purchase of large tracts of farmland, Los Angeles Water and Power held water rights to much of Owens Valley. Water exported from the valley enabled Los Angeles to grow into the present-day megalopolis. By 1930, the city owned 95 percent of Owens Valley farmland, and a fifty-mile stretch of the Owens River had dried up. By 1948, many pupfish pools were nothing more than shallow bowls of cracked earth. By 1953, when Phil joined the ranks of the California Department of Fish and Game, researchers had put this creature out of mind as irretrievably lost. But in 1964 biologists made a thrilling discovery: a few Owens pupfish were still alive in a place called Fish Slough, another site of Phil's work, in the center of volcanic tableland in Owens Valley.

On that Sunday morning, Phil and I drove Highway 6 toward the reverse islands of Fish Slough, a series of spring-fed pools at the base of the jagged Sierra Nevada accessible only by washboard back roads. Reverse islands like those found in Fish Slough depend entirely on a network of springs boiling upward from the belly of the world, flows that have circulated through the Earth's veins for 8000 years or more before surfacing. As Phil negotiated the road, he explained that, although hundreds of aboveground miles separate these waters, a complex labyrinth of aquifers snake underground like the chutes of a water park and form the Death Valley hydrographic region. Overpumping from any single site within this network may pinch off the flow to a spring miles away.

As Phil described what's at stake for preserving the unique habitat of Fish Slough, his brows turned down and the lines on his forehead, visible beneath the bill of his cap, crowded together the way lines on a topographic map suddenly bunch when elevation rises steeply. As we approached the spring where Phil rescued the pupfish on that August day in 1969, he said, "There was no 'how-to' manual for preventing the extinction of a native species. My colleagues and I had to figure it out on our own."

The discovery of a residual population prompted Phil and a handful of other conservation-minded government managers to work during their free time for native fish, while "making fishing better for the hordes of visitors from L.A." when on the clock. They fought during the 1960s to protect Fish Slough and the Owens pupfish against further onslaughts. While telling me the story of the Owens pup, Phil quoted Aldo Leopold's *Round River*, "One of the penalties of an ecological education is that one lives alone in a world of wounds." He explained that without the backing of the Endangered Species Act, which didn't pass until 1973, managers like Phil, who favored native species over introduced ones, felt solitary and largely unsupported. "But we needed to do it," he said. "It was the right thing to do."

When Phil and I arrived on foot at the spring, he spoke about that afternoon in 1969, just as biologists celebrated the fifth anniversary of the Owens pupfish resurrection. On that August day, Phil's colleague Bob Brown scuttled breathlessly into the office. Phil mumbled something to the receptionist, and he, Bob, and John Deinstadt sprinted out the door into the glaring sun, clutching nets and white five-gallon buckets. According to Bob, the Owens pupfish were about to go extinct—again.

Three unforeseen factors had converged to set in motion the second near-extinction of the Owens pupfish. As Phil explained, the winter of 1969 was fierce and brought unusual amounts of rain and snow to the desert, which led to astounding growth of vegetation that spring. A phenomenon known as evapotranspiration followed. Temperatures crested 100°F with no humidity, causing the levels of evaporation to be high. Transpiration, the normal metabolic process of plants taking up water from nearby pools, streams, and ponds, was also unusually high due to the amount of new vegetation. Phil explained it to me this way: "Imagine a saucer of water with six people sitting around it with straws. Then suddenly the number doubles. The pools and streams where the pupfish lived were depleted quickly." On that day in 1969, some of the last Owens pupfish in the world had already turned belly-up to the sun.

Phil and I stood staring into the spring surrounded by tall, golden shafts of tule stalks. "How did you net the fish back then?" I asked. "Did you wade into the spring with your nets, or scoop them from the ledge?"

"We didn't have to wade at all," he answered. "The spring had dried to the point where there were only a few puddles remaining." He and the two other men could walk into the middle of the spring and stay dry in a pair of street shoes.

Phil's impulse to preserve desert fishes stems from his belief that a pupfish is no more and no less deserving of life than a human being. For Phil, the inherent value of pupfish outweighs any potential usefulness they have to people. But when asked, "What good are pupfish, anyway?" he keeps in mind that we live in a time when few voices question human values. He answers that some of these beings inhabit water many times the salinity of the ocean and may aid medical science in understanding such things as kidney function. Other times, Phil replies from his belief in the value of aquatic life for its own sake rather than for any benefit to humans. He returns the question, "What good are *you*?"

As Phil told me the story of that day, I imagined him and the other men gunning down rutted dirt roads, buckets knocking. In that sweltering heat, they must have been dizzyingly hot and thirsty, layers of dust clinging to their sweat as they plunged their nets into the muck.

His voice grew more passionate as he continued the story, "I distinctly remember being scared to death," he said of netting 800 of the pups from the shrinking pond and transporting them to another spring. He'd been guided by geneticists' estimates that at least 200 fish were needed to maintain a viable species. The three men, knowing they would lose some of the fish to severe stress, scooped blindly, netting every fish they could. "Lugging a full bucket in each hand, I realized I held the existence of an entire vertebrate species," he said. "If I'd tripped over a piece of barbed wire or stepped in a rodent burrow, the Owens pupfish would be extinct."

I pictured Phil weighted down on either side, lugging those buckets across a desert surface ruptured by heat and littered with dry, twisted branches of dead and dying shrubs. He held in his hands the life of a species about to blink out.

In the back of the pickup truck, Phil transported the fish to a nearby spring. That's where Phil and I headed next, climbing back into the white Suburban to bump and jounce down the same route he'd taken that August day. As we navigated the dirt track, Phil's hands gripping the steering wheel, I saw him as he must have looked thirty-odd years ago, driving as fast as he could while avoiding any troughs or holes. Those men must have held their breath, cursing each time they hit a bump. They would have craned their necks for a look at the buckets, eyes scanning for any fish sloshed out and frying on the hot metal truck bed.

At the spring, we ducked under the wire fence and stood on the spot where Phil had released the pups and watched them night and day for signs of success or more death. As we gazed into the spring, I caught a grin creeping over Phil's face. He seemed lost in thought for a moment, perhaps remembering some detail from those days. Because we visited in winter, there were no pups gamboling about, but I felt their presence, the steady pulse of their tiny lives in the mud. And from Phil's face, I know he did, too. "They'll be out again in spring," he said.

Enough individuals survived that the Owens pupfish still exists, and biologists today repeat Phil's first acts of preservation. Resource managers transplant Owens pupfish from one spring to another, mixing genes to keep the species hardy and to maintain backup populations in the event of a die-off. But they fight an uphill battle against ongoing threats to pupfish populations. Ash Meadows refuge manager Sharon McKelvey, with only three staff members to manage more than thirty springs and pools, orders special crawfish traps and baits them with cat food to catch the larger adults. The problem, she said, is that crawfish breed continuously. And these mudbugs aren't the only problem. Mosquito fish, intentionally introduced in the 1900s by local governments as a malaria-control agent, inhabit pupfish pools, eating more pupfish eggs than mosquito larvae. "In fact," Sharon said, "the pups eat more mosquito larvae than the mosquito fish do."

California Governor
Edmund "Pat" Brown
discusses conservation
strategies with Phil
Pister (right) in 1959.
Photograph provided
by Phil Pister

Another significant source of invasive species is pets released into the pools by people who think they're doing a nice thing by setting their frogs, turtles, and aquarium fish free. "People aren't malicious. They just don't know the damage their pets can do." A lot of released pets don't survive, but enough do to become dangerous to the pupfish. Two of the biggest pet threats to pupfish are the red-eared slider, a popular pet turtle, and the convict cichlid, a Central American fish commonly sold for aquariums. The convict fish grows up to six inches in length and is not a community fish, meaning it can't share an aquarium with any other fish because it will eat them. This fish spawns continuously, and the refuge's three employees can't seem to trap them fast enough. McKelvey echoed Phil's opinion that, short of a miracle pot of money to hire a private company to eradicate the most invasive species, her staff can just manage to keep the pupfish populations at healthy numbers despite the invaders. "Restoration is spring by spring," she added.

What she said next mirrored the sentiments of every island manager I spoke to. "We're trying to tip the balance back in favor of native species," she said, sounding optimistic but tired. "Our goal is to make these invaders a smaller component within a mostly native community." But it's hard to do without money. As Phil put it, "Funds are never adequate to fulfill the intent of the Endangered Species Act when it comes to actual expenditures for species recovery. This problem has been more acute in recent years because of the priorities of Republican administrations."

As part of a formal species recovery plan that Phil developed, such detailed and careful acts of preservation provide evidence of Phil's achievements as a conservation biologist. Phil even has a pup named after him, *Cyprinodon pisteri*, Pister's pupfish, native to northern Mexico. When I ask Phil if he's had other creatures named after him, he says, laughing, that in fact there is a snail, *Pyrgulopsis pisteri*, "named by a guy back at the Smithsonian." In response to the Owens pupfish crisis and the explosion of residential growth in the arid Southwest, Phil and others founded the Desert Fishes Council, which is dedicated to the conservation of North America's arid land ecosystems.

Although much of Phil's career was a life-and-death game of beat the clock, youthful enthusiasm for the preservation of desert fishes lingers in his voice, while his knowledge is that of a veteran biologist. According to Phil, he fuels his work by communicating with interested nonscientists. "Writing a scientific paper on pupfish is valuable," he said, "but I really try to get my colleagues in academia to look at preservation a little more pragmatically. Unless we can get the conservation word out to the public, who ultimately make decisions through the political process, much of our work's value is lost." To that end, and following his retirement from the California Department of Fish and Game, Phil writes for general readers in magazines like *Natural History* and lectures to audiences all over North America and the United Kingdom.

Crystal and I piled into the car and drove to another site of Phil's conservation efforts, Death Valley National Park, thirty miles west of Ash Meadows. We wanted to see pupfish living in colder creek water. The Salt Creek pupfish lives in the waters of Death Valley's Salt Creek, where cold water stretches its life to two years, instead of the average one-year life span of its cousins.

When we arrived, we stepped onto a boardwalk that serpentined out of sight through the dunes edging the creek. Signs along the pathway discouraged visitors from climbing the fragile drifts, and I imagined mastodon bones buried beneath them. Peering into the winding and shallow brook, I glimpsed a silvery Salt Creek pupfish. Seasonal evaporation had created water-filled depressions separated from the stream's main channel and corralled several pupfish.

Crystal and I sprawled on our bellies to study the stream's contents. "I know we're not supposed to do this," Crystal said, dipping her fingertips in the stream, "but I just have to know how cold it is for the fish." She shook her reddened fingers and laughed, "It's cold."

Chins in our hands, we watched several pupfish wriggle around in the inch-deep water. While some of the fish may die when these temporary sinks dry up from time to time, a stray rivulet sometimes runs between the puddle and the creek's main channel, whisking a fish along its watery bridge to safety. Pupfish frisked and gamboled. The males, hostile toward others of their sex, darted from their hiding places and head-butted each other, giving chase through a jungle of sedges.

A family approached the spring with a girl of about two or three years old in tow. Occasionally, a careless parent might permit his child to lob rocks at the fish, but this little girl seemed gentle, fascinated, and as giddy as Crystal when she put her fingers in the chilly water. When the girl squatted next to me, she leaned forward to wag her pudgy index finger in the water. An inquisitive pupfish swam toward it, and the girl shrieked, yanking both hands to her chest and stamping her feet on the wooden planks. "Fiss," she said to me, pointing at the stream, and I smiled at her. Her quick movements and high voice sent the slender shaft of the pupfish back into hiding. But he'd be back.

The wind picked up, and grains of sand swirled and heaped along the edge of the boardwalk and caught in our teeth. Water boatmen, belonging to the order of insects known as Hemiptera, or true bugs, skated on the pond's surface. They zipped and bounced between the ragged, shrinking edges of the puddle like bumper cars on a rink. As Phil explained, Salt Creek pupfish wait for the boatmen, one of their favorite foods, to die so that they might eat them, the same way vultures watch for animals to straggle down to their deaths.

I watched one pupfish and then another catch a wave back to the main artery of Salt Creek. I thought of Phil Pister with his buckets and something he'd said to me, "The most self-interested thing humans can do is to show concern for other species. Because ultimately it comes down to preserving the habitats we all occupy and depend on." I leaned over the boardwalk and dipped my palm into the vanishing aquatic island to ladle a pupfish and place him back in the creek.

Chapter 3 Poor Man's Galápagos

Trying to remember the Gulf is like trying to re-create a dream. This is by no means a sentimental thing, it has little to do with beauty or even conscious liking. But the Gulf does draw one. . . . We know we must go back if we live, and we don't know why.

—John Steinbeck, *The Log from the Sea of Cortez*

Jacques Cousteau called this Mexican sea "the aquarium of the world." El Gulfo de California, the official name used by the Mexican government since the early twentieth century, suggests that this ecologically rich body of water is an appendage of the state of California. Vermilion Sea, while poetic, is a name gone largely out of fashion; and as names go, it says little about the place. The Sea of Cortés is a name derived from the Spanish soldier and invader Hernán Cortés, a conquistador who embodied both endurance and cunning when he massacred the Aztecs. But he met his match in one of the most arid regions of the Americas—the Baja Peninsula and its offshore islands. I have to admit I prefer the name Sea of Cortés because it conjures up the memory of how the peninsula vanquished the great conqueror and his men with disease, ambushes by native people, and a shortage of water and food.

The Sea of Cortés began to form about 5 million years ago, when the San Andreas Fault opened, ripping a finger of land 800 miles long from what is now mainland Mexico. The Pacific Ocean slowly flooded the tear in the earth, but left several mountain peaks uncovered. Exposed crags bunched around the seam of the fissure are the Midriff Islands, sandwiched between the central Gulf Coast region of Sonora, on mainland Mexico, and the central Baja coast along the slim waistline of the Baja Peninsula. Often referred to as the "poor man's Galápagos," these islands are accessible to a person of lesser means, whereas the true Galápagos Islands, located 600 miles off the shores of Ecuador, are very expensive for North Americans to get to.

There are hardly any indigenous people left on the Baja Peninsula. Although many native plants and animals still thrive on the peninsula and

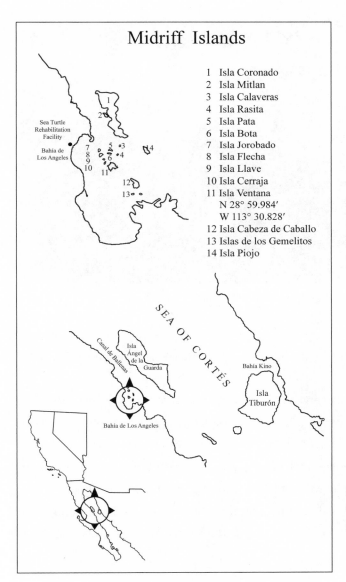

Midriff Islands

1 Isla Coronado
2 Isla Mitlan
3 Isla Calaveras
4 Isla Rasita
5 Isla Pata
6 Isla Bota
7 Isla Jorobado
8 Isla Flecha
9 Isla Llave
10 Isla Cerraja
11 Isla Ventana
    N 28° 59.984′
    W 113° 30.828′
12 Isla Cabeza de Caballo
13 Islas de los Gemelitos
14 Isla Piojo

The Midriff Islands
in the Sea of
Cortés. Illustration
by Rick Moser

A view of the Midriff Islands in the Sea of Cortés

on the arid Midriff Islands, invasion by exotic species threatens their continued existence.

From California, my husband Bruce and I will drive down the Baja Peninsula along Mexico Highway 1, *The Grebe* strapped to the roof of our pickup truck. We're headed to the eastern central Gulf Coast, to the village of Bahía de los Angeles, a logical departure point for invasive species dispersing to the Midriff Islands. In their essay, "Invasive Plants: Their Occurrence and Possible Impact on the Central Gulf Coast of Sonora and the Midriff Islands in the Sea of Cortés," Patty West and Gary Nabhan listed six exotic plants that pose the greatest threat to these islands: crystal iceplant, tamarisk (*Tamarix chinensis*), Russian thistle, buffelgrass, giant reed, and Sahara mustard.

Each of these six species has a fascinating history of entrance and spread into Mexico and the western United States. Some, such as crystal iceplant, are still actively planted. Although all six invasive plants are high priority for island management and native species conservation, I chose to pay particular attention to crystal iceplant, tamarisk, and Russian thistle, as well as another species of iceplant known as the Hottentot fig, because I encountered these plants most frequently on my journey to the Baja Peninsula and the Bahía de los Angeles archipelago, and they appear as common invaders on the other islands explored in this book.

These species share some common traits, including extreme salt and drought tolerance and the ability to spread easily once established, often assisted by people, birds, waterways, and wind. Another common denominator for these invaders is their habit of establishing and spreading in disturbed soils. As we reach the increasingly arid regions of the peninsula, Bruce and I will see which of these invaders have made it as far south as Bahía de los Angeles. From there, we'll paddle to the small cluster of Midriff Islands just offshore to see if we detect any invasive species there. According to interviews with biologists studying the Sea of Cortés and a few written accounts, plant and animal invaders haven't yet established robust populations on the Bahía de los Angeles archipelago, unlike the Baja Peninsula and many of the other Midriff Islands, where cats, rats, house mice, buffelgrass, and tamarisk pose significant threats. We're going to see for ourselves.

With *The Grebe* mounted on the ladder rack, nylon straps cinched tight over the kayak's bow and stern, we drive California Highway 101 through Santa Barbara and Ventura, where I come upon my first Hottentot fig, commonly referred to as "highway iceplant" or "dead man's fingers." The plant is an aggressive species native to South Africa, and it carpets the entire coastline with mats of three-sided succulent leaves. Crowned with bright yellow and pink flowers, Hottentot fig is so prolific along the California coast that author Elna Bakker noted that "Few postcard stands lack pictures . . . of Mesembryanthemum [crystal iceplant], a South African genus that has donated a number of colorful species now so common to coastal bluffs that they are usually thought to be native."

The California Department of Transportation (CALTRANS) originally planted iceplant species on coastal bluffs to prevent erosion and along railroad tracks to stabilize soils. CALTRANS continues to plant Hottentot fig along California's highways because once established, these plants require almost no tending. I spoke with a woman from CALTRANS District 7 who asked that I not use her name. "Some people think it's a problem plant. It is. It grows anywhere," she said. "Iceplant is our coastal frontier plant," she went on. "We plant it in Camarillo, Ventura, and all along Highway 101 and the Pacific Coast Highway. It grows so low to the ground that it doesn't block ocean views, and we plant it wherever there's new road construction and highway expansion."

The Hottentot fig, still planted along California highways, is now growing in Bahía de los Angeles.

A CALTRANS architect, who also declined to give his name, warned "If a runner of the Hottentot fig goes down a storm drain, wherever it comes up there's a good chance it will establish a new colony." However, another district landscape architect with CALTRANS in San Diego told me, "It's a favorite with our maintenance teams because it requires so little care." He explained that planting of Hottentot fig first took place in the 1930s, primarily for dune stabilization, ground cover, and as a fire retardant because it doesn't burn well. "Our limited maintenance budget really drives what species we choose to plant," he said, "and iceplant is self sustaining. It's a good plant. It has its place."

CALTRANS maintenance workers simply cut a two-foot-long runner

Crystal iceplant has spread throughout California and is appearing along the Baja Peninsula.

of this species from an existing plant and bury one foot of it in the ground where they want to establish a new colony. Within a month the plant takes root at the nodes and spreads. This same district landscape architect suggested that if my home had a large yard, especially property with a slope, iceplant would make a nice groundcover. "You can buy it at any Home Depot or nursery," he added.

Crystal iceplant, the other major invasive iceplant species, reproduces a little differently than the Hottentot fig. According to Nancy Vivrette of the Ransom Seed Lab in Carpenteria, California, crystal iceplant produces many seeds to ensure reproduction. Humans and birds are the most likely agents of dispersal for this species. With seeds stuck to their feathers and

Carpets of Hottentot fig actively planted for ornamental and erosion control purposes

feet, birds may carry the seeds from place to place. West and Nabhan maintain that birds and pack rats use iceplant for nesting material.

"However, seed germination is a function of soil disturbance," Vivrette said. If there is no disturbance, seeds will lie dormant, viable for twenty years or possibly more, and native vegetation will keep the seeds from germinating. But if there is a natural or human-induced disturbance that ruptures the integrity of a native plant community, such as road construction, housing developments, or increased foot traffic, iceplant seeds will sprout, and as Vivrette put it, "This plant will climb up and over whatever native vegetation surrounds it," smothering out the native plants it shadows.

Bruce and I stop at a selection of nurseries in Ventura and Oxnard to

Iceplant species for sale in a California nursery

verify the sale of iceplant. Sure enough, flats of various species, including Hottentot fig, are available for $12.99 a flat without any warning about their invasive characteristics. Marcy, a helpful employee at one of the nurseries we visit, slides the *Sunset Western Garden Book* from beneath the cashier's counter, saying that this is the book most nurseries consult and avid gardeners keep handy on their bookshelves. I find the page describing iceplants: "Where hardy, they are among the most useful and colorful of flowering ground covers. . . . All [iceplants] take [to] most soils." This ability to grow in a variety of soils allows iceplant species to establish and flourish in any climate similar to its South African home.

Bruce and I connect to Interstate 5 via Highway 126 toward the international border. The lines of cars at *la frontera* stretch half a mile, and carpets of Hottentot fig and crystal iceplant roll out of sight on both sides of the border. We'll wait here for some time before beginning the long haul down Highway 1, the slender ribbon of pavement that runs the length of the Baja Peninsula. While letting the truck idle in line, I wonder just how far south this plant has managed to spread.

Thin men with squeegees wash tourists' windshields. Someone cuts in front of someone else, and horns blast. People get out of their cars to wave police to the scene, and the rule-breaker is sent to the back of the line. As we creep nearer the concrete arches separating the two countries, on the Mexican side I see old women barely more than four feet tall, with round bodies and head scarves tied under their chins, selling something.

Once across the border, we're officially on the Baja Peninsula. One of the women leans against the driver's door and presses her cardboard box of Clorets and Halls through the open window. I buy two packets for ten pesos. We tail a group of Mexican men packed into a Honda Civic. When we reach a stoplight, I half expect a hatch on the roof to pop open and one man after another to climb out like the twenty-seven clowns who emerge one after another from a tiny clown car in the circus. The driver steers along Tijuana's flooded streets, where a broken fire hydrant has showered the scorched earth and made the pavement slick. We take a hard left and glimpse the Civic's driver ahead of us peering through his windshield cracked in the pattern of a spider's web.

With Bruce at the wheel, I study the map, taking special note of land-marks along the borderlands and Highway 1, known as the Transpeninsular

Highway. When naturalist Joseph Wood Krutch took this road to Bahía de los Angeles in the 1950s, it was a rutted, often washed-out dirt track. He packed sixteen gallons of gasoline for the trip, a smart move considering the numbers of tourist cars we see upside down in hedges of cholla cactus: gutted and stripped Ford trucks and rusted Volkswagen buses, wheels to the sky like turtles that couldn't right themselves and died belly-up.

After Mexican authorities paved Highway 1 in 1973, Krutch wrote about his conflicted feelings concerning increased tourism and development on the peninsula: "Have I any right to feel a twinge of regret? Certainly not if the hotel means a better existence for those whom tourists will bring what will no doubt be, in some respects at least, a more abundant life. Perhaps I should rejoice also to think that more citizens of the United States will have an opportunity to visit the magnificent scene. But of this last I am less sure."

The Midriff Islands are still relatively insulated from tourism pressures, but this is rapidly changing due to an avalanche of large development projects on the peninsula intended to attract visitors and foreign investors. Such development projects degrade habitat and provide the soil disturbance needed to facilitate seed germination of crystal iceplant and other invaders.

We'll see what kinds of development projects are currently underway on the peninsula. Meanwhile, my finger traces several completed development projects along the border. The Morelos Diversion Dam, on the Arizona-Sonora border, allows U.S. cities and agricultural projects to siphon off

Tamarisk spreads rapidly once established.

95 percent of the Colorado River's flow, while a scant 5 percent reaches Mexico. Irrigation canals that trickle water into the Mexicali Valley appear like spidery veins on the map. Exotic plants use these canals and rivers to spread into areas previously inhabited only by native plants. Spotting clumps of tamarisk along the Tijuana River, I get Bruce to pull over just after we cross the bridge. The waterway is congested with Styrofoam and the most notorious tamarisk species, *T. chinensis*, no doubt benefiting from disturbance to the Colorado River's natural volume and course.

As early as the 1820s, nurseries in the United States offered this Eurasian plant for sale. By 1900 the Arizona Agricultural Extension Service encouraged tamarisk planting for landscaping because the plants grow rapidly and provide agricultural crops with protection from wind and soil erosion. However, intentional planting of tamarisk became a terrible irony. The same farmers who sought this tree's attributes didn't realize that the plant had a nasty habit of thieving water, not only from native plants but from their own agricultural crops. Tamarisk escaped from its original planting sites and began spreading wherever human disturbance occurred, most often above and below dam sites.

The economic costs of tamarisk invasion are significant. According to Erika Zavaleta of Stanford University's Department of Biological Sciences, the western United States alone loses 1.4 to 3.0 billion cubic meters of water to tamarisk each year. Over a fifty-five-year period, she estimates that California and Arizona alone will experience a loss of between $1.4 billion and $3.7 billion as a result. This figure doesn't include the loss of agricultural water supplies.

Ecological costs of tamarisk invasion are also high. In a single year, one tree is capable of producing 500,000 seeds, which wind and water disperse over long distances. Tamarisk can prosper in a variety of habitats. It flanks irrigation canals, chokes wetlands, and clogs ponds and springs. Its latticework of roots trawls the soil for every drop of water, drying up springs and displacing riparian woodlands made up of native trees, such as cottonwood and willow and the other vegetation the endangered south-western willow flycatcher relies on for foraging and nesting.

Zavaleta's research supports her assertion, shared by many biologists, that the costs of tamarisk invasion and attempts at controlling it make a strong case for preventing invasive species introductions in the first place. "When faced with a decision either to spend billions of dollars in control or to tolerate tens of billions of dollars in continued damage by the invader, one is reminded that both costly choices could have been prevented—and could be prevented in the future—by decisions to prevent introductions at an earlier stage." Tamarisks also have been replacing native riparian species because dams have halted the annual floods that traditionally created seedbeds for native plants. So, even if tamarisk is removed, unless annual floods are allowed to recur, no native species will repopulate the area.

A tamarisk tree twice my height grows near the road along the Tijuana River. Its attractive pink flowers cluster together at the end of each branch and bob in the wind. I can see why people imported the tree as an ornamental. It's beautiful. However, salt excreted from the tree's foliage dusts the ground near my feet, thwarting the germination of any native seeds that are not salt tolerant. I break off a branch and put my tongue to the jointed stem. It tastes like the ocean. After establishing in drainages across the arid and semiarid climates of the western United States, tamarisk has found its way into northwestern Mexico. As I stand above the canal wondering how much tamarisk I'll encounter along the peninsula, the bloated carcasses of two dogs float by.

About twenty miles north of Ensenada, I pull the passenger seat back into an upright position, having drifted off to sleep. I focus on the abundance of coastal sage scrub—fourwing saltbush and compass barrel cactus. Along the disturbed edges of the highway, exotic Australian saltbush grows low to the ground and in abundance.

We blaze through Ensenada, with its Blockbuster, Starbucks, Auto Zone, and Day's Inn. Just south of town a fifty-foot plaster Jesus lies prostrate on a hillside above the highway. The figure's glossy paint gleams in the sunlight. I notice Bruce's foot press harder on the gas pedal. "We're outta here," he says. "That's just weird."

In the town of El Rosario, we pull over for crab tacos and a Pacifico at Mama Espinoza's. Outside the restaurant, a plant that grows in my mother's garden, commonly known as garden nasturtium, sprouts the edible tuba-shaped orange flowers my mother plucks for her summer salads. Native to Peru and introduced to the peninsula in the 1700s, nasturtium carries the infamous label "alien invasive," and its ornamental and edible qualities are the plant's only saving grace.

We pull the truck into the Pemex station up the road for the last petrol available for several hundred miles. El Rosario is considered the end of civilization by many Baja travelers and the entrance into a raw and undeveloped peninsula. There we'll be lucky to encounter a man selling liters of petrol from the back of a pickup truck. For now, we'll try our luck on a full tank and a five-gallon can in the bed of the truck. With the tank filling, I wander the station to stretch my legs and there it is—the Hottentot fig with its yellow flowers.

As I stand here, arms crossed over my chest, nose twitching from gasoline fumes, I wonder if this patch of iceplant will be the last I see, or if there's more beyond this point. Probably, I think to myself. Nancy Vivrette mentioned visiting the peninsula in the 1960s and seeing patches of crystal iceplant in disturbed areas. Images of crystal iceplant, with its pearly bladder cells making it look perpetually wet, and pickle-shaped Hottentot fig climbing up and over every native plant in their way, rewind and play in my head.

Vivrette had told me that crystal iceplant was likely introduced by ships trading in the Sea of Cortés and along the Pacific Coast, docking in harbors and inadvertently carrying the seeds in cargo containers onto the peninsula. Trade as a factor in the spread of invasive species certainly didn't begin nor end with crystal iceplant. Globalization of the economy, like that provided for in the North American Free Trade Agreement (NAFTA) and the United States–Dominican Republic–Central American Free Trade Agreement (US-DR-CAFTA), is also responsible for aiding the spread of invasive species.

Anne Perrault, project attorney at the Center for International Environmental Law, collaborated with researchers from Defenders of Wildlife on a paper addressing the role of global trade in spreading invasive species. Presented at the Second North American Symposium on Assessing the Environmental Effects of Trade, held in Mexico City in 2003, the paper indicted global trade as a major contributor to the spread of invasive species: "While impacts of invasive alien species are primarily local and national, the root causes of their spread are regional and international—driven primarily by global trade, transport, and tourism."

There's an endless supply of horror stories. Packaging used to transport goods internationally is often the Trojan horse for invasive species. Hilary French, director of the Worldwatch Institute's Globalization and Governance Project, cites the 1998 ban imposed by the United States on the import of Chinese goods packed in untreated wooden crates after discovering their role in introducing the Asian long-horned beetle, an invasive insect that tunnels through hardwood trees, girdles trunks, and eventually kills the trees. French also cites ballast water as a major contributor to populations of nonnative aquatic species, "On any given day, some 3000 to 10,000 species are moving around the world in ship ballasts. When the ballast water is discharged, so are the organisms."

The zebra mussel is one example of an invasive species first released into the Great Lakes around 1985, along with ballast water from the Baltic Sea. Native to the Caspian Sea, this fingernail-sized mollusk now chokes the waterways of eastern North America, where it consumes copious amounts of algae, a major component of aquatic food systems. But shipping traffic doesn't simply facilitate a one-way invasion. While over thirty of the more than 100 exotic species living in the Great Lakes are Baltic Sea natives, a similar number of species native to the Great Lakes now reside in the Baltic Sea.

Author Christopher Bright noted in his article "Invasive Species: Pathogens of Globalization" that when placed into historical context, the occasional introduction of a nonnative species is natural, and an ecosystem can likely handle the invader. However, an ecosystem and its inhabitants cannot absorb the radical species-swapping caused by accelerated and unmanaged global trade. The real problem, in other words, does not lie with the exotic species themselves, but with the economic system that is continually showering them over the Earth's surface. Bioinvasion has become a kind of globalization disease.

Policymakers have not yet integrated the language of bioinvasion into their discussions of trade policies, and the negative consequences of invasion do not inform economic laws. Bright suggested that expansion of global commerce without careful consideration of environmental impacts is shaping evolutionary processes in ways we cannot even foresee, and any effort to slow the rate of spread breaks down in the face of dominant policies and economic interests that favor unfettered trade.

Energized by Mama Espinoza's crab tacos, Bruce and I decide to drive as long as there's still enough light to dodge oncoming semis with their homicidal drivers. Appearing as a gentle curve on the map, the Transpeninsular Highway is actually a series of tight curves with no shoulders and steep drop-offs into ditches. One truck driver, who we decide must be on maximum-strength No Doz or maybe PCP, passes us on a sharp uphill curve where it's impossible to see oncoming traffic. We get lucky. There happen to be no cars coming from the opposite direction—this time—but the truck driver could not have known. If there had been oncoming traffic, we might all be dead. "That guy's nuts," Bruce shouts. "We need to get off the road until morning."

We pull off the road and sleep in the open truck bed. By midnight the temperature drops and tiny ice crystals form on my sleeping bag. In the morning, I wake in a garden of plants that look like upside-down parsnips and raise myself on my elbows to look around. I flip through the pages of Norman Roberts's *Baja California Plant Field Guide*. The upside-down parsnip is actually the species *Fouquieria columnaris*, or *cirio*, named for the tapered candles of Mexico's Catholic missions. This tree has another, perhaps more memorable name than either its Latin or Mexican title. In 1922 science and literature united when botanist Godfrey Sikes applied the name "boojum" to this singular species, after a word in Lewis Carroll's "The Hunting of the Snark," the longest nonsense poem in the English language.

It seems as if all the native plants have assembled here and the exotics I tallied yesterday were part of a bad dream that included Starbucks. The vegetation, the dips and rises of the landscape, are completely unfamiliar, but everything belongs here—except us, of course. I worm out of my sleeping bag and waggle stocking feet into mud-caked sneakers.

We're at the mouth of the Vizcaíno Desert region. The change is sudden, as if a knife had etched an indelible line between the coastal scrub of the far north and this inland, drier vegetation. For all the strange life forms growing here, we could be at the center of Earth or on the surface of Jupiter's moon Io, for all I know. I feel like Lewis Carroll's Alice sliding down the rabbit hole, with plenty of time to glance at the strange surroundings and wonder what might happen next. When I press my nose to the cream-colored skin of the *cirio*, I smell a mixture of carrot and ginger, musty and sweet.

Navigating toward the interior of the peninsula, we see fewer invasive plants. Native plants dominate the vast expanses of uninhabited land that stretch between Mexican ranches. Here and there, houses of clapboard and tin appear with clusters of tamarisk surrounding their walls. But where little or no soil disturbance has occurred, native plants such as chain-link cholla grow in abundance. This plant's green fruit forms links that easily snap off, and a new plant will root where the fruit falls, a refreshing example of native plant reproduction.

Just north of Cataviña, along the main artery, we gape out the window at a forest of *cardón* trees. These columnar giants are leafless. As true cadophylls, their green stalks do the work of photosynthesis, producing necessary carbohydrates for energy.

To the uninitiated, the Baja Peninsula is stark and beguiling, but romanticizing Baja comes easily when survival isn't in question. We look at peninsular life from the point of view of visitors who can easily come and go. We don't loathe the place for what it can do to a human body unable to find water or shelter. In his 1866 account of the Baja Peninsula, J. Ross Browne expressed the depth of his revulsion: "Every shrub armed with thorns; the cactus, in all its varieties, solitary and erect, or in twisted masses, or snake-like undulations, tortures the traveler with piercing needles and remorseless fangs. . . . You have a combination of horrors that might well justify the belief of the old Spaniards that the country was accursed by God."

Not alone in his grousing about the Baja, Browne spoke for nearly every explorer who visited the peninsula. They all found it deadly. People starved and died of thirst. As late as the mid-twentieth century, the peninsula wore down even the most adventurous who tried to penetrate its interior. Those who made it out alive returned home slender as a clam's neck. They could have learned something from the Cochimí, who mined their community dung heap for undigested *pitaya agria* seeds, then toasted and ground them into flour for cakes and a mealy mush to subsist on. The Spanish were repulsed at this survival technique, calling it the Cochimí's "second harvest."

Bruce and I dodge cows and follow the highway's switchbacks, peering into dry washes where the plumelike branches of tamarisk stick up like cowlicks. Pulling over at the turnoff to the old onyx mines at El Marmol, I jump out to pee behind a crowd of Russian thistle stuck to the barbed wire fencing surrounding a Mexican ranch. The wind picks up, and balls of this spiny plant roll into my legs and poke my backside.

Not a true thistle but a chenopod (a member of the harmless-sounding goosefoot family), Russian thistle barbs nevertheless stick to my boots and shins and bloody my fingers as I gingerly pluck them one by one from my shoelaces and underwear. Russian thistle was introduced to the Midwest around 1870 in a grain shipment from the steppes of Mongolia. Around the beginning of the twentieth century this tumbleweed appeared in the American Southwest, where it was dubbed "white man's plant" by the Hopi. Russian thistle, which breaks from its roots as a mode of seed dispersal, became as much an icon of the Wild West as the cowboy myth,

Young
Russian thistle
growing in
dense thickets

the mustang, and the Duke. Hollywood cast the tumbleweed, also called "wind witch" and "witch weed," in supporting roles on screen, and The Sons of the Pioneers and Roy Rogers ennobled this portable brier patch in the famous range lullaby, "Tumbling Tumbleweed":

> See them tumbling down
> Bearing their love to the ground . . .
> Here on the range I belong
> Drifting along with the tumbling tumbleweed.

Another piece of tumbleweed and then another clings to my socks as

Mature Russian thistle breaks off at the base of the stalk and tumbles along the landscape, spreading its seeds liberally.

I pick my way back to the car, and I'm reminded of how Russian thistle changes from the romantic tumbling tumbleweed to an invasive exotic. In summer and autumn, the prickly shrub dries out and waits for the first hot winds to barrel through. The wind snaps the head off its stalk, and the thorny ball rolls across the landscape, dispersing its seeds, insuring the plant is well represented among the flora of the region. This seed-dispersal mechanism is no accident, but an intricate, tried-and-true evolutionary design. It works in its native habitats, and it works particularly well in its nonnative range, where there are few mechanisms to check its spread.

Bruce and I swap seats again, and I take the helm. But first I scour the tires for any seeds or root bits of exotic plants clinging to the tires of our pickup. There they are, crammed into the rubber's pattern, just waiting to pop out midway down the peninsula. With a bandana wrapped around my fingers, I wheedle the seeds from the rubber, place them in a plastic bag, and stash them in the cab. While there's only a small chance that the seeds would germinate, the increasing rate of tourism to the region puts the odds in favor of Russian thistle and other invaders reaching the islands, unless active management and public education are stepped up.

As we enter the area of the peninsula once inhabited by the Cochimí, Russian thistle fades from my attention and native old man cactus appears. Its gray spines crown the top of each succulent arm growing skyward from a basal cluster. Its night-blooming flowers open on spring nights to court endemic moths that pollinate the flowers and lay their eggs inside them.

Among strange life forms and invasive plants, signs announcing Escalera Nautica, or "Nautical Stairway," stake the highway, heralding the coming of prosperity to the peninsula. With money from foreign investors, the project will include an eighty-mile-long road paved through a forest of boojum trees from the Pacific Ocean to the Sea of Cortés, saving yachters the long journey around the cape, the southern tip of the peninsula. The Mexican government is optimistic that boaters will entrust their multimillion-dollar vessels, with names like *Second Wind*, *Success*, and *Summers Off*, to local hauling services that will cart them to Bahía de los Angeles, a village slated for the construction of one of twenty-odd luxury marinas replete with spas, hotels, and golf courses.

FONATUR, Mexico's tourism promotion agency, pitches the project to Baja residents as a major economic boon to their villages. However, I had spoken with another traveler, Jeff Watkins, a retired community college instructor from Oregon who owns a vacation home south of Bahía de los Angeles near Múlege. He described abandoned luxury hotels in Napolo and Puerto Escondido, two other projects resulting from a joint effort by the Mexican and French governments. "Puerto Escondido has a working port," he said. "There's an anchorage, a boat launch, and storage facilities. The rest of it is crumbling, including a quarter-mile-long dock." Watkins also mentioned a boarded-up extravagant restaurant and condos that no one ever stayed in.

Pro Peninsula, an environmental group based in San Diego, California, doesn't believe Escalera Nautica will materialize. But the group cites all of Mexico's failed development schemes on the peninsula as major culprits in habitat fragmentation that allow invasive species to take advantage of soil disturbance and replace native and endemic plants. Environmental impacts have never been considered when implementing such schemes, the group said.

Two miles outside the village, Bruce and I stop at Programa Tortuga Marina, a loggerhead sea turtle rehabilitation site. Antonio Resendiz founded the program in the 1970s. When we arrive, we find Antonio cleaning the turtles' blue nylon holding tanks. He dresses their wounds with a bright purple antiseptic solution.

As we watch the turtles stretch their necks toward water streaming from a hose, I ask Antonio what he thinks about Escalera Nautica. "Sí, the Road to Hell," he says. "If people don't know it already, they'll eventually figure out it's a sham." Antonio waves a rubber-gloved hand at the grounds surrounding the turtle facility. "Look around you," he says. "Escalera Nautica isn't the first attempt to turn the peninsula into a tourist playground. It's just another development scheme that will fall apart midway."

The rehabilitation center sits on the site of an abandoned RV park. Tree yuccas line rows of RV sites never used. Gutted buildings, with Russian thistle balls wedged where windowpanes have long since been knocked out, stand like ghostly monuments to failed and expensive tourist infrastructure—like a party no one attended. "These failures just chip away at the habitat integrity of the peninsula and the islands. Even when they don't go through, look at what they leave behind," he adds, shaking his head.

Antonio drags the hose to another tank and tells us we should really talk to Guillermo Galván, the owner of a restaurant in town. "He'll have a lot to say about this development business."

Bruce and I sit outside of Guillermo's, a restaurant overlooking the bay and the islands we'll paddle to first thing tomorrow. Galván is visiting a neighbor, his daughter says, setting down two Tecates and a basket of oily tortilla chips cradled in wax paper. A retired, eighty-nine-year-old miner named Herman pulls his chair up to our table. A Californian by birth, Herman tells us that he decided to retire to the bay and has been a resident of this stretch of seacoast for the last nineteen years.

"What do you call four Mexicans drowning?" he says, adjusting his hearing aid and leaning forward.

"I give up," I say.

"Cuatro sinko," Herman says, grinning.

Dogs skulk around the restaurant, their backbones shifting under loose skin. I feed them greasy chips, although I know I shouldn't because feral dogs and cats often harass and kill native birds, reptiles, and mammals. Dalmatian, shepherd, and every other breed appear in various combinations. The males have their testicles, and the females are pregnant or nursing, their bellies swollen, teats dragging. Some dogs scratch their mange sores until they break open and bleed. Others sleep in the shade of athel trees (*Tamarix aphylla*), another species of invasive tamarisk intentionally planted around houses for windbreaks and shade.

Eyeing me while I feed the dogs tortilla chips, Herman remarks that nobody in the village gets their dogs fixed. "These animals sometimes go out to the islands with the fishermen," he says. "Cats, too. They let them loose out there and they go feral. People are lazy, letting their male dogs and cats run around with their balls on. They're the easiest to fix—just tie a rubber band around their balls. I put anti-freeze out for the cats—makes them demonstrably dead." The dogs probably don't stay on the islands, I think to myself, but if they defecate on the islands, they could introduce nonnative seeds.

Herman tells us that a few years back, in response to complaints by the people of Bahía de los Angeles that too many feral dogs prowled the village, the Mexican Army showed up in town and told residents that any dogs they wanted alive should be locked inside their homes. The army rounded up more than sixty dogs roaming outdoors and shot them, heaved the carcasses into the back of a dump truck, and drove away.

When Guillermo Galván returns to the restaurant for lunch, we excuse ourselves and join him. With unregulated development, Galván explains, "We villagers will have to give up our lives as fishermen to become waiters and janitors, maids and caddies. Where's the dignity in that? And we don't want our islands ruined."

Galván says that if the people of Bahía de los Angeles get to share in the profits generated by new hotels and restaurants, then the project may be a good idea. It's a long way to the middle of the Baja Peninsula, and with little water and no gasoline available, only the most dedicated traveler ventures beyond El Rosario de Arriba and Cabo San Lucas. "We want more tourists to visit here. We want more income and basic necessities, but we don't want new businesses to overrun the place and steal the few tourists we do get."

He tells us that he wrote a letter to Mexican President Vicente Fox, expressing these concerns. Galván requested that the president order the new marina and hotels to be built in Bahía de los Angeles, but on a smaller scale, with villagers maintaining local control over not only tourism development but also island management. In his letter, Galván also asked the president to consider the long-term good of peninsular residents. Residents want more of the basic necessities, like running water and a sewage system. Presently, water and other supplies have to be trucked into the village once a week, making everything but fish expensive.

"How did the president respond?" I ask.

Galván shakes his head, "I never got an answer."

Bahía de los Angeles is not inundated by invasive species. They are present and many are poised to invade the islands, but the village and its archipelago provide an incredible opportunity for preemptive action. Every biologist I've spoken to emphasizes prevention. It's easier to thwart introductions of exotic species than to get rid of them once they've established. According to ethnobiologist Patty West, the next best action is to eradicate invasive species while their populations are still small and to consistently monitor the islands for new colonies of exotic plants. When I spoke to West, she emphasized the need for involvement of local people. But right now, she said, Baja residents can't easily get permits to remove invasive species, and added, "Scientists don't get down there enough to observe changes to island vegetation and spot new invasions. A few local people with knowledge of plants should be authorized to remove any invasive species they observe."

West also suggested establishing a system for island visitors to report any nonnative species they observe to help managers identify potential threats early. "People go to these places because they think they're beautiful. They want to help preserve them," West said. She added that local workshops in Spanish about the impacts of invasive species would help, and the common boat-launching areas in the village should be monitored for invasive plants to help prevent them from dispersing to the islands. "It takes a lot of awareness," West said. "Even scientists forget that they might carry invasive seeds on their socks and shoes, and they go from island to island. It's a long process of developing awareness and changing habits. Pulling seeds off your socks and checking your gear are helpful."

From the beach, where crystal iceplant has formed a thatch over the dunes, I walk to the police station to register for a permit to land on the islands. A step toward managing island access, this registration system is similar to the one implemented by the National Park Service in the United States. Visitors go through a short orientation to the islands, including where to camp, and pay a small fee that goes back into the community and into island conservation.

Wandering down the village streets, I spot Hottentot fig growing luxuriantly in several yards and clothes drying on a line strung between two tamarisk trees. I resist the urge to pounce on these private plots, yanking the succulent runners, clawing out the cantaloupe-colored roots, which is one effective eradication technique. Unlike *T. chinensis* and *T. aphylla*, crystal iceplant and Hottentot fig don't spread by seed. These latter species would have to be carried to the islands—not an unlikely scenario and one worthy of monitoring. Crystal iceplant has an edge here. The climate and conditions are similar to its native South Africa, with high salinity and a short winter rainy season followed by drought. It's the perfect habitat.

That night on the beach, I lie near a copse of athel, with one of the skinny town dogs nestled next to my sleeping bag. Feeling my shoulder bones against stone, loose pebbles beneath my knees, I hear coyotes yip somewhere in the desert. How many, if any, invasive species will we encounter on the islands tomorrow, and which ones? The weight of moonlight on my eyelids forces me to sleep.

The next morning, up with the gray light of dawn, Bruce and I lug provisions to *The Grebe*, including nine gallons of water. The skeletal dog slinks off to look for breakfast. While loading the hatches, I notice a garland of scratches on the kayak from landings on rocky island shores. Like scars with a good story attached to them, they ring the glistening hull as we pack dry bags of clothing and food, stuffing them in the front and rear hatches until sweat stings our eyes. We tug neoprene booties over our feet, tighten the Velcro waistbands on our spray-skirts, and shove off, pointing the kayak's nose toward Isla Ventana, our first island.

We ease our strokes as we pass an abandoned fishing boat, the *San Agustín III*, pirated by brown pelicans. Every inch of space along the boat's railing is crowded with the birds, and the boat's paint has been worn away or burned off by their acidic guano. We drift alongside the fishing boat and cast a line with one of the Lucky Joe lures that Herman had given us, hoping to catch either a mackerel or some other sizeable fish for dinner. After an hour with a slack line, we begin to think that we have no gift for fishing and give up.

Then we hear the piercing *kayoo-kayoo-kayoo* of one of the greatest fishers of all, the osprey. The bird captures our attention, and we look skyward to see the osprey struggle to maintain its grip on a fish. While boobies and brown pelicans dive bill-first for their quarry, the diurnal osprey plunges feet-first, clutching a prize that is always arranged headfirst in its talons. Brown pelicans, two-thirds the size of the American white pelican, harass the osprey, attempting to pilfer its meal. In the tussle with the pelican mob, the osprey loses its grip. The fish, whose scales give off the cold silvery light of tinsel, whips its tail side to side before slapping the sea surface and sinking out of sight. After another attempt at another fish, the osprey, successful this time, flies to the top of a *cardón* to eat in peace.

Distracted by the osprey, we don't notice the slight breeze nudging the kayak's nose off course, toward Isla Cabeza de Caballo, Horsehead Island. In March 2000, scorpion researcher Gary Polis of the University of California, Davis, along with three visiting ecology professors from Japan and a postgraduate researcher, died after spending the afternoon studying Horsehead Island's scorpion population. On the return trip, less than four nautical miles from Bahía de los Angeles, winds gusted and whipped the sea into gravity waves, large waves capable of traveling thousands of miles without breaking. The other type of wave, capillary waves, would have come first, like a person's breath blowing on a hot cup of tea, creating a rippling effect on the surface but not disturbing the skiff as Polis and the other passengers headed for the shelter of the bay. Then the wind whisked the sea to froth, creating four- to six-foot gravity waves that swamped Polis's boat and flipped it over. As we watch brown boobies plunge for fish and picture each coming day of paddling and observation as if it were a perfectly wound clock, Bruce and I don't know that this will be our calmest day on the sea, the only day without any danger.

Heermann's gulls flash their bright red bills and emit a throaty *huh-huh-huh*, as we glide into a crescent-shaped inlet and glimpse a stone arch on the south side of Isla Ventana, a window in the rock for which the island, the second largest in this archipelago, is named.

A couple hundred yards ahead, we see a possible landing site and nose the kayak to shore. The tide is low, and the kayak's bow runs aground on sharp rocks. Bruce tugs upward on the loop of his spray-skirt, and we scramble out of our cockpits to lug the boat out of the surf. We'll have to trundle across exposed algae-covered rocks and heave the boat above the high-tide line. Bruce wades to the bow and attaches a line. Skidding on a mat of algae, he falls forward into the surf and skins his knee. "Shit. Watch it, honey. It's slippery out here."

I wade to the stern of the kayak and slide the boat forward, lift and slide, lift and slide, to reduce scoring along the hull by saw-toothed barnacles. We unload the center and rear hatches, tossing water bladders, food bags, cooking pots, field guides, and sleeping gear to the back of the beach on dry ground. By the time we get *The Grebe* above the high-tide line, we both have raspberries on our knees and elbows, and my teeth are clattering loudly.

The first thing I notice is a rusted metal billboard, taller than my near six feet. The sign, in Spanish and English, instructs island visitors not to collect plant material and to camp only in designated areas. Another sign, in pictures, illustrates prohibited activities: a human hand attempting to pet a sea lion, boaters ferrying dogs and cats to the islands. The others represent the usual island prohibitions—no campfires, no trash left on the island, and no making off with the cacti and other plants. The signs are a joint effort of the Secretariat of the Environment, Natural Resources, and Fisheries; the Arizona-Sonora Desert Museum; and local residents, who erected these low-cost, low-maintenance signs on all common landing and camp sites, even constructing stone pathways on some island trails to encourage visitors not to traipse all over the island's fragile vegetation.

Isla Ventana is designated as an Area Protección de Flora y Fauna. Some Mexican activists are advocating for the entire Sea of Cortés and its islands to be categorized as a biosphere reserve under Mexican federal law. But for now, this island and the fifteen others clustered around it have at least some legal protection.

Reading the signs in our dripping shorts, we squat in the sand to wrestle off our booties. Bruce disappears down the beach, gathering a couple flat stones as he goes. People always leave traces of their doings, even behaviors we can't control, like defecation. Environmentalists and ecologists still know no better way to dispose of human waste on islands than to make what they call "shit pucks." Considered the most sanitary way to defecate on a coastal island, this technique requires a visitor to scout two flat rocks and a private site. The bottom rock catches the dung, while the

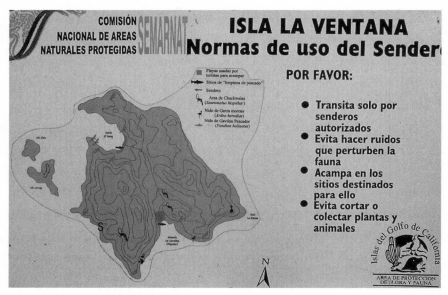

A conservation sign located at Isla la Ventana landing site

top stone is used to squash the whole mess together. The visitor walks to the waterline and shot-puts the puck as far as he can muscle. This keeps intrusion to a minimum.

When he returns from the shoreline, Bruce finds me grazing on native pickleweed, a briny, tart green the size of my pinky nail that has crept along the dunes, forming an impenetrable hedge along the island's beachfront. I roll the jointed stem with my tongue, the taste a great deal more salty than sour.

Then my gaze falls on it, a thatch of crystal iceplant covered in what look like ice crystals, giving it a perpetually wet appearance. These blisters are actually bladder cells that sequester and store salt away from the rest of the plant's cells. As Vivrette explained, the seeds, every one of them fertile, self-pollinate and can withstand a good dunking in seawater. In fact, the seeds can float in their buds, rafting from island to island.

We hike a short canyon trail lined with a species of native fourwing saltbush called *cenizo* by the Mexicans. The succulent leaves of the saltbush are fat with recent and unexpected moisture. There had been rain a few days before we arrived, an unusual event for this area. Higher up, along the dry, windy uplands of the island, indigenous jojoba bushes dominate the landscape. This is the plant whose oil, which is actually a liquid wax, was used by indigenous people of the Southwest to treat dry, chapped skin and rashes. Jojoba is often listed on shampoo and conditioner products.

The wind from the north has sheared Isla Ventana's jojoba shrubs, rendering them naked on one side, although branches facing the island interior support an abundance of yellowish green, leathery leaves. It's possible, however, that this half-and-half model is a conservation strategy of the plant. Other jojoba shrubs on islands farther north have similar patterns of leaf distribution. Perhaps the plants, aside from being windblown, leafed out on only one side because the cost in energy to maintain the windward leaves is too high. Water in any form is a currency more precious than black pearls in this place, and successful invasive plants in arid lands often possess greater drought tolerance than the native plants they displace.

Backtracking across the island to our camp, we catch sight of a fifteen-foot-tall concrete cross on the highest point of the island, overlooking the Sea of Cortés to the east. Like most of the sea's islands, Isla Ventana supports no permanent human communities. Yet, every island displays a cross and evidence of human visitation. In the 1700s, when missions were established successfully on the peninsula, the padres instructed the Cochimí to erect a cross on the highest point of each island to repel evil.

Ocotillo sprouts a rush of green leaves along its branches following the recent rains. Its long, ropey limbs tower four feet above the crown of my head. Ocotillo looks like the kind of living thing that might have inspired Dr. Seuss. Streaks of green run the full length of each arm, and these striations conduct the plant's photosynthesis when it's not in leaf. The tree's tubular crimson blossoms grow along the tips of each branch and are favored by the Costa's hummingbird. I have never seen more than one hummingbird at a time before becoming acquainted with this ocotillo. Now I have witnessed a charm of hummingbirds, as an aggregation of them is known. But this tree offers more than hummingbirds; native people make a strong root-and-flower tea from the plant to relieve the wet cough of an old person.

By nightfall the wind settles, and we unpack the boat hatches to set up camp near the fish-cleaning station, a *sitio de limpieza de pescado*, used by Mexican fishermen. A neatly arranged wall of rocks crowned with skeletons of giant manta rays and unidentifiable big-headed fish with buckteeth serves as our windbreak for cooking. We crawl into our bags with our heads against the rock wall.

Sometime during the night, I'm startled awake by the feel of clammy skin against the backs of my hands and the sound of something scratching in the dirt near my head. I feel a light tug on my hair. The creature disappears into the pickleweed when I bolt upright. Maybe I've been dreaming. Asleep on a waterless island, surrounded by the sound of lapping seawater, I might have dreamt of clammy moisture. My mouth is as dry as cotton balls, and sand grinds between my teeth. I suck down half a liter of water before going back to sleep. When we wake, evidence of night visitors lies in our breakfast bowls. We rinse the brownish black pellets of the native Merriam's kangaroo rat from our dishes and fix oatmeal and grainy coffee.

After breakfast, poking around the littoral zone, the area between high and low tide lines, we find a native jumbo squid approximately three or four feet long. I recognize the squid from a specimen I had seen months earlier at the Santa Barbara Museum of Natural History. The squid's triangular fin flutters against smooth wet pebbles along the shore. The tide animates the glutinous body, kneading it backward and forward like bread dough.

The fishermen of the Sea of Cortés call this squid *diablo rojo*, red devil, for its ability to turn from white to red in the blink of an eye. They tell stories of men snatched by the crushing parrot-shaped bill of this fierce, carnivorous feeder and dragged to the sea bottom. Although the specimen in front of me isn't fully grown and is probably harmless out of the water, I'm not about to investigate more closely.

We leave the squid and climb to the rocky ledge overlooking the beach, where I peer through binoculars in the direction of the peninsula.

Glimpsing the tiny village of Bahía de los Angeles, I remember something naturalist Joseph Wood Krutch wrote in *The Forgotten Peninsula*, "The worst features of the primitive are unhappily combined with all the shabbier features of sophistication. . . . If Los Angeles Bay manages to get the fringe benefits of 'progress' without paying very much for them, they may just possibly be an ideal condition."

Determined to see chuckwallas, iguanas that live on the Midriff Islands, we skulk through saltbush and peer into rocky crevices. These lizards bask in the sunlight until their bodies reach an optimal temperature of 100°F, then they withdraw beneath boulders and into clefts to avoid overheating. Two spiny chuckwallas sun on flat stones. Alerted by the vibrations of footfall, the chuckwallas scurry into crevices, where they suck air, inflating the baggy folds of skin hanging around their necks and torsos until their bodies wedge snugly between stones. When they bloat their bodies in this way, it's difficult for predators to catch them. Indigenous people who ate chuckwalla meat didn't give up easily. They extricated the lizards by stabbing them with sharpened sticks to puncture their bellies and let out all the air. Then they tugged the lizards from their hiding places and roasted them. We press our faces into the shadowy cracks for a peek at their ballooned bellies and then leave them alone.

When we return to the beach, we find western gulls scavenging around the fish-cleaning site. The wind has picked up, what seamen call "light air," a 1 on the Beaufort scale. Ripples and scales characterize the waters in the cove, but there are no whitecaps visible. We estimate wind velocity at one to three miles per hour and decide to put in for Isla Coronado (also called "Isla Smith" on some maps). I recall a passage from expert kayaker Andromeda Romano-Lax, who wrote, "Gazing seaward from your campsite, safely nestled in a cove or bay, you may not be able to see how the waves and wind are clashing just around the next point. Expect conditions to be rougher than they appear." We load the boat hatches, poking at a pelican skull washed in by the night tide and no bigger than a goatnut, and then shove off for Isla Coronado. At four and a half miles in length, it is the largest island in the bay. As we near the western side of Isla Ventana facing Isla Cerraja and Isla Llave (Lock and Key Islands), the wind grows stronger, but we decide that it's still a gentle breeze, eight to twelve miles per hour, with wavelets cresting over the bow.

Bruce and I coordinate strokes, steering the kayak's nose into the oncoming curls to avoid broaching. In the event we do capsize, we can access the bilge pump strapped to the topside of the kayak and bail seawater from the cockpits. The rock bottom is suddenly visible, sheeted by orange starfish and jade-green plants. Then, just as abruptly, the bottom disappears again and the waters of the strait quicken—an ocean beltway, sweeping us around large, submerged rocks. We jounce and jar along this conveyor belt for a couple hundred yards. I recall something Steinbeck wrote in *The Log from the Sea of Cortez*, "The Sea of Cortez . . . is a long, narrow, highly dangerous body of water."

Our goal is to reach the northernmost cove of Isla Coronado to see if we observe any invasive plants there. But these winds appear to be gaining in strength, pushing *The Grebe* backward. Then the spray starts. I bow my head to keep my glasses from streaming. This must be what Romano-Lax means when she says that winter paddling in the Sea of Cortés requires "extra flexibility" with itineraries. Bruce wants to push on. "We still have to round the southern tip of the island," I say, craning my neck to shout to him. "It's probably going to get worse. Let's put in."

Although both of us are disappointed by a short day on the water, we also know beaching is a wise choice. With each stroke, gusts catch my oar and nearly jerk it out of my grip. By the time we reach the shore of Isla Pata, we're just thankful we haven't capsized. The waters encircling the peninsula are known for their winter winds, the "Baja bash," that blow the full 2200 miles of peninsular coastline. Paddling into the clear, shallow channel between the small Isla Bota (Boot Island) and Isla Pata (Paw or Foot Island) is like exiting a wind tunnel and suddenly entering silence. My ears hum from the wind's boxing as we beach the kayak.

Peeling off our wet suits, Bruce and I watch as waves climb and the wind quickens. By late afternoon, whitecaps roll into the lagoon. Isla Pata has a friendly crushed-shell beach, and its well-established pickleweed hedges protect our camp from some of the wind's force. The flora isn't complex. Smaller islands require less taxonomic effort to identify plant and animal life, but small islands with fewer species are also more vulnerable to disturbance (usually human disturbance) and loss of native plants due to invasion by exotic species. I don't think any chuckwallas live here, but that doesn't take away from the excitement of this place. Scrambling to a ridge-

The Grebe on the shores of Isla Coronado (Isla Smith), where the author
set up camp

top, I spy an osprey nest measuring two feet by three feet. There are no
occupants this time of the year, so I don't hesitate to crawl along the crumbling neckline of the island and peer at the nest's construction of sticks and
feathers, human hair, a doll's leg, bones, and shreds of toilet paper.

We go to bed early, prepared for a predawn launch, skipping coffee
and breakfast the next morning to make a swift departure. We look forward to reaching Isla Coronado, an island with a larger array of plants and
animals due its size and its greater variety of habitats. After listening to

waves lunge at the island all night, we know it will be rough going. Bruce and I stuff our mouths with miniature Baby Ruth bars and dig into the surf until our arm muscles burn. Tightening our abdominal muscles, we bow our heads and lean into the spray. Two porpoises surface for air about ten yards away. Not more than four feet long, they may be *vaquitas*, the shy endangered and endemic porpoises whose numbers have dwindled due to loss of breeding grounds and overfishing of one of their favored prey fish, the endangered *totoaba*. Willing to sacrifice a pair of inexpensive field glasses for the possibility of seeing a rare bird or marine mammal, we tug on the straps of our binoculars, buried inside our lifejackets, and try to focus the glasses on one of the small cetaceans. Gazing through lenses smeared with saltwater, I can just make out the diagnostic grayish black stain around the eye, a shiner that makes the porpoise look as though it's been in a bar fight.

The porpoises disappear and our attention reverts to keeping the kayak on course. Isla Calaveras (Skulls Island) and Isla Piojo (Louse Island) lie to the east, and we know that as we round the southern end of Isla Coronado and head due north, the work will get only more difficult and dangerous. There are no other boats on the sea this morning. Even boobies and pelicans remain in the safety of lagoons and beneath rock overhangs to wait out the squall.

As we clear the southern tip of Isla Coronado, a six-foot wave slams the port side of *The Grebe*, causing the kayak to heel, to lean too far to the right. "Heave to," I yell back to Bruce, who's in the rear cockpit. He maneuvers the rudder to help us put the bow straight into the wind and keep the boat upright. We have to stay far enough from exposed rocks at the base of Isla Coronado's guano-covered cliffs to keep from breaking onto them, which would crush the kayak and probably us too. Stroking in this weather is like shoveling wet cement. Winds gust up to thirty-five miles an hour, and the sound deafens. Our hats skid to the backs of our heads, kept on by well-cinched chinstraps that dig into our throats. The fetch of the wind, the distance over which the wind blows without a change of direction, seems endless, and the waves hurtle us to shore on a diagonal.

For safety, we decide to camp on this beach for the night. As I begin to unload the rear cockpit, my hands poised to haul out the tent, I hear Bruce call to me through the gusts of wind that drub my ears. I crane my neck to see him standing in wet sand, pointing at the high tide line running along the base of the sheer cliffs that climb out of sight behind him. If he hadn't noticed the high-tide mark, we'd have woken in the middle of the night with the surf in our sleeping bags, the kayak washed out to sea, and no dry ground to escape to.

Lips tinged blue, even though we're outfitted in full-body wet suits, we resolve to get off the beach. I clamber in first and seal my spray-skirt, while Bruce steadies the boat from the stern. He prepares to push us off the beach and leap into the rear seat, when a wave hits the kayak broadside and causes it to yaw, to swing right and left. The bow slams into a barnacle-covered boulder jutting out of the surf, and I push against the rock to unwedge the hull. It's no good. Out of the corner of my eye, I see Bruce struggle to keep the boat upright, but sand and surf flood the open rear cockpit, making the kayak unwieldy. As another breaker enters shallow water, it does what oceanographers call "feeling bottom." Contact with the sea floor slows the bottom half of the wave, while the top half spills over it with force and turns the kayak's hull to the sky. Just before my head goes underwater, I hear Bruce yell, "Pull your loop!" But my spray-skirt is secure around the coaming, and no matter how hard I pull on the safety loop, the neoprene cloth won't budge.

I'm trained to exit a capsized boat at sea. Usually people simply fall out due to gravity. But I have no training in how to avoid drowning while stuck on the beach. Flood tide fills my nose. My left arm is trapped under the boat. People drown in shallow water. I can't drown, I think to myself, we're supposed to reach the northern tip of this island. I'm here to look for exotic species. I've got work to do. A mixture of fear and indignation takes over, and I press my forehead against the hard-packed sand. With my right hand, I wrench the loop of my spray-skirt. This time it gives, popping off the rim, and I snake out from beneath the boat on my belly, my mouth full of the Sea of Cortés.

Fear of what this wind can do fills my body. The pain of burning muscle has no place to hold fast as we point the bow into the waves and launch successfully from the beach. We stroke across the swells, waves known to ancient mariners as Kymothoe, described in the *Iliad* as "the wave that has traveled a great distance."

With our heads bent against the spray, we glide farther out into Canal de Ballenas (Canal of Whales) to get away from the breakers. We just catch a glimpse of Isla Coronado's volcano, which rises 1500 feet out of the sea, when the boat suddenly weathercocks, swinging left and then right and back again with the action of the wind. We turn the bow south, toward a protected cove, but now instead of us paddling against the sea, the sea is chasing us. Called a "following sea," the waves are both dangerous and thrilling the way they overtake the stern of the kayak, lifting the boat several feet in the air and setting it down again, the surf washing down our backs. We coordinate our movements to keep the kayak from capsizing and carefully edge it toward the beach. After several minutes, a wave dumps us on the sand.

We drag soaked sleeping bags, pads, and sopping field guides from the hatches and spread them out on the smooth, pebbly beach at the base of the volcano. I lay out the books, separating chunks of pages with stones to allow the air to save what it can, including my copy of Steinbeck's *The Log from the Sea of Cortez*.

To help us forget the sodden field guides and gear, Bruce and I climb halfway up Volcán Coronado and perch on chunks of andesite to share the bottle of wine we'd been saving, the label now unreadable and peeling away. That night we sleep in a sandy depression below the volcano, in wet but warm sleeping bags, protected from the wind.

In the morning, the sea now calmer, we paddle to the southernmost end of Isla Coronado and peer into clear lagoon waters that harbor a garden of red sea urchins and Cortés round stingrays, with a wingspan of two feet, hidden in the sand and silt of the sea bottom. We also see a smack of moon jellies and a Cortés halibut, a flat fish with both eyes on the same side of its head, located close to a slightly upturned mouth with protruding lips in a permanent pout. On the beach looms the familiar rusting billboard we've encountered on the other islands, prohibiting visitors from removing plant material, introducing plants and animals, and dumping garbage.

Through binoculars I have a clear view of Isla Ángel de la Guarda, where shipwrecks and fishermen accidentally introduced Norway rats and house mice. Later, their island camps raided by both, fishermen attempted to rid the island of native and exotic rodents by intentionally introducing cats, now considered the most invasive mammal on the Midriff Islands. Ecologists such as Bernie Tershy of Island Conservation and Alfonso Aguirre of Grupo Ecología y Conservación de Islas, among others, work diligently to eradicate invasive species on the islands wherever possible.

But today, on the south end of Isla Coronado, my attention is on crystal iceplant, which conceals several dune areas. There is no method currently known for eradicating this plant once it's established. Maintaining native habitat integrity is the only way to prevent it from spreading.

From our campsite, we hike a well-trodden trail over the cactus-covered hills to another inlet, where we find a red mangrove (*mangle rojo*) thicket growing with its tangled roots in the marsh. These native shrubby trees appear like a single entity rather than several individuals tangled together. The knotted trunks, thin as stilts but all brawn, anchor the trees in the mud and allow them to grow upward. Slight branches support elliptical leaves placed on opposite sides of the stem; greenish white star-shaped flowers are in bloom. Seeds of the mangrove will germinate while still attached to the parent tree. After germination, they drop off and float until they contact mud, where they establish and grow in other shallow lagoons with good tidal flushing. Mangrove forests are nurseries for snails, clams, small fish, and other sea life, providing them protection from rough seas and predators. Some fishermen call these forests "the roots of the sea" because they are one of the most fertile habitats on Earth, supporting a generous variety of life.

Bruce and I nod to a group of kayakers nestled near the mangrove swamp. This lagoon is extremely fragile, and yet the area is frequented by campers and should perhaps be closed to camping as tourism on the island increases. Foot traffic around the mangrove forest and across the island may disturb soils enough to allow iceplants and other invaders to establish. One of these is buffelgrass, a species recently added to Arizona's noxious species list. Native to southern Africa, buffelgrass was originally introduced to the Southwest by the U.S. Department of Agriculture Soil and Conservation Service in the 1940s. Sonoran and Texan ranchers are still planting buffelgrass for cattle feed. I spoke to Bruce Eilerts, head of the Arizona Department of Transportation's Natural Resources Program. "After researching our records and talking with career employees, I've found no evidence that we ever planted buffelgrass," he told me. Eilerts, along with John Hall of The Nature Conservancy and Larry Riley of Arizona Game and Fish, pushed an executive order to address the problem of invasive species in Arizona, including buffelgrass. In 2005 Arizona Governor Janet Napolitano passed Executive Order 2005-09 and established the Invasive Species Advisory Council. The council's mission is to "develop a consensus vision for a coordinated multi-stakeholder approach to invasive species management in Arizona." Eilerts emphasized that stakeholders include everyone affected by invasive species: federal and state agencies, Native American groups, agricultural interests, and even pet interests, among many others. Along with sowing native seed mixes along the state's highways, the Arizona Department of Transportation is now considering the formulation of an eradication program, according to Charles R. Barclay, Natural Resources Management Section manager in Tucson.

These steps are being taken none too soon. Although successfully eradicated from Isla Tiburón, buffelgrass has established hardy populations not only in southern Arizona, Texas, and northern Sonora, but also on Isla Alcatraz of the Midriff chain. Its wind-dispersed seeds can withstand saltwater immersion, according to West and Nabhan, making it a significant danger to the Bahía de los Angeles archipelago.

Patty West also told me that *T. chinensis* has established a colony on Isla Alcatraz. Its wind-dispersed seeds can establish in saline and brackish soils and require only the slightest amount of fresh water to germinate. Known for its tenacity, this invader can thrive when inundated by saltwater. The dunes surrounding the red mangroves would be vulnerable to tamarisk because, while seemingly dry to the casual observer, they soak up rainfall and harbor heavy layers of moisture that the tamarisks' deep roots can readily access.

The dunes are also vulnerable to invasion by the crystal iceplant, because this species doesn't need fresh water to germinate, thriving on salt water alone. In fact, seamen carried this plant aboard ship because they could water it with the ocean, boil the leaves, and eat it like spinach. While iceplants do prevent or slow erosion of dunes, they also restrict the natural movement of sand necessary for a healthy dune system.

Although campers disturb fragile soils, making them more susceptible to the wind-dispersed seeds of invasives, West doesn't advocate closing the mangrove swamp to visitors. She does agree that moving the campground to a less vulnerable location and increasing visitor education about habitat degradation and the negative impacts of invasive species would be a good conservation move. "People want to help," she said. "They just need to understand the reasons for island regulations." If people do understand, she believes, most will comply with island policies. Every island visitor I've spoken to is eager to know more about invasive species, validating West's optimism.

At dusk, near the campsite, we crouch behind a hedge of pickleweed to observe a year-round resident of these islands, the reddish egret. This shy bird, known for its comical feeding behavior, lands on its long legs in the lagoon and begins its fishing dance. It jigs, pirouettes, and pounces for fish in the shallow mud flats, wings suspended overhead to cast a shadow over the water and more easily spot its prey. How many reddish egrets will be able to do their dance here if the mangrove nursery of its food source is impaired?

During our third night on the southern end of Isla Coronado, we hear crickets in the pickleweed and see the cinder cone black against the emerging stars. We've camped on four islands and cruised along the shores of many others, and the tally of invasive plants has been exactly two hardy patches of crystal iceplant. It doesn't seem like much, but there may be more here than we were able to detect. Tomorrow Bruce and I will paddle back to Bahía de los Angeles, where other invasive species are poised to make the voyage we have made. As we watch brown pelicans fly over us in the dark, the wind in their wing feathers emitting a metallic whistle, we know that invasive plants on the other side of the gulf await a strong wind to carry their seeds to island soils. The moon rises cold like a wedge of Granny Smith apple, illuminating the campsite and the lagoon, where stingrays camouflage in sand and mud, their disk-shaped bodies hidden.

As we watch fish jump occasionally in the lagoon, Bruce recites a few lines from the poem "La Luna Asoma" ("The Moon Rises") by Spanish poet Federico García Lorca:

Nadie come naranjas	Nobody eats oranges
bajo la luna llena.	under the full moon.
Es preciso comer	One must eat fruit
fruta verde y helada.	that is green and cold.

Perhaps the Cochimí drew such a moon on peninsular rock. The moonlight reveals the bleached leg bones of an animal strewn around our campsite.

That night we sleep with our backs to the bones. I don't learn until we return to the village that two women float dead on the Sea of Cortés a few nautical miles from our camp. Two women and a man had taken a leaky boat with a five-horsepower engine—less than the kick on my food processor—and only one lifejacket among them. When afternoon winds gusted and waves pitched, the boaters couldn't bail water fast enough. The man, brother to one of the women, ended up with the jacket. He alone survived. Our moon must have shone on the drowned women's skin, gleaming like the flesh of green apples.

Chapter 4   The Pied Pipers of Anacapa

Save the Pigs!
—Aerial banner in the sky over Santa Barbara, California, April 2005

Trailing behind a biplane, the red-lettered banner decries the killing of feral pigs on Santa Cruz Island, the largest of California's northern Channel Islands. On the street below, protesters from the Channel Islands Animal Protection Agency (CHIAPA) condemn the pig eradication scheme as merciless, a waste of taxpayer money, and based on bad science. In front of a placard that reads, "Do these pigs look like a threat to the environment? Of course not," demonstrators occasionally exchange kisses with one of two Vietnamese potbellied pigs they have brought as mascots to the rally.

Meanwhile, the DJs of a popular local radio station offer a prize to the caller with the best alternative to the $5 million eradication project. The whole event turns into a kind of carnival when one caller suggests relocating the pigs by helicopter or ship to a farm somewhere, while still another listener alleges that the science behind the project is flawed. The pigs, she says, don't hurt the island's native wildlife in any way and should be left alone. Another caller suggests contraceptives to slow the rate of reproduction. Someone else proposes that the National Park Service and The Nature Conservancy, the sponsors of the pig-eradication project, drop two or three of the feral pigs into the backyards of every person protesting the killing. "Let's see how nice a pet those 500-pound sows make," he says. The prize offered by the radio station for the best idea is an autographed IPOD containing every song ever written by the rock band U2.

Dubbed the "Channel Islands" because they are separated from mainland California by the Santa Barbara (or Chumash) Channel, this chain of arid and semiarid islands is home to plants and animals found nowhere else in the world. The four northernmost islands, San Miguel, Santa Rosa, Santa Cruz, and Anacapa, are oceanic islands, never connected to mainland California but joined to each other approximately 18,000 years ago when sea levels dipped to their lowest levels. As sea levels rose, water flooded the valleys between the larger peaks and divided the 724-square-mile superisland, called "Santa Rosae," into four individual landmasses. Unique life forms evolved on each of the four islands, plants and animals so isolated that the Channel Islands, which are managed mostly by the National Park Service as Channel Islands National Park, occasionally go by the name "Galápagos of the north." Their miles of undisturbed coastline are inhabited by breeding birds, California harbor seals, sea lions, and even whales.

The story of the Channel Islands shows that many people do care about islands enough to rehabilitate those that have suffered centuries of degradation. The drama being played out on the islands today, however, is likely to repeat itself wherever bureaucratic decisions strike an emotional chord in nearby communities. What is happening on the Channel Islands is a striking example of the conflicts that can rage when eradication of a charismatic mammal is proposed. The leading players in the drama are, predictably, a government agency that isn't entirely candid with the public and its antagonists, concerned citizens and animal-rights groups who sometimes have more passion than accurate information.

Ninety-six square miles in size and the largest of the eight Channel Islands, Santa Cruz, "Island of the Holy Cross," or as the Chumash Indians called it, "Limuw" (meaning "in the sea"), could also be called "Scorpion Island." Its body is shaped like a scorpion, tail poised to sting.

Bruce and I take turns driving Highway 101 along the coastal range toward Santa Barbara, where we plan to catch a concessionaire's boat to Santa Cruz. We stop only occasionally to tighten the straps on *The Grebe*, which is riding on the roof of the truck. Our goal, to kayak among the northern Channel

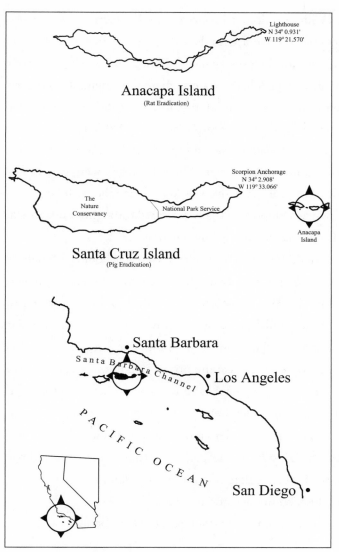

Lighthouse
N 34° 0.931'
W 119° 21.570'

Anacapa Island
(Rat Eradication)

Scorpion Anchorage
N 34° 2.908'
W 119° 33.066'

The
Nature
Conservancy

National Park Service

Anacapa
Island

Santa Cruz Island
(Pig Eradication)

Santa Barbara

Santa Barbara Channel

Los Angeles

PACIFIC OCEAN

San Diego

California's
northern
Channel Islands.
Illustration by
Rick Moser

Islands, will take us paddling along the seaway of the Chumash, who maintained a thriving culture on Santa Cruz Island for more than 10,000 years and depended on the sea for their sustenance. Although we won't rely on our fishing prowess to eat, we'll paddle the waters where the Chumash navigated their *tomols*, wooden plank canoes crafted from local pine or hewn from large drifting hunks of redwood from northern California. With tools made of stone, shell, and bone, the Chumash sanded these planks to a smooth finish and made them watertight with sharkskin. The tribe nearly went extinct following the 1965 death of the last native speaker of the Chumash language, and people with Chumash blood are largely scattered throughout the coastal cities of southern California, including Santa Barbara.

We roll into Santa Barbara and drive along Cabrillo Boulevard, named after Juan Rodríguez Cabrillo, a navigator of uncertain national origin who served under Hernán Cortés and directed a battalion of crossbowmen to defeat the Aztecs. Bruce and I get a hotel room downtown and wander along State Street, passing shops and restaurants like The Bar with No Name and The James Joyce. We're careful not to step into the suicide bike lane without looking both ways twice, and then one more time for good measure—not for fear of bicycles but of speeding cars. We split up to explore different shops, and I meander along side streets into Ortega Park. Murals painted on the outside of the park's restrooms and outbuildings turn my attention to the particulars of Santa Cruz Island colonial history and how it shaped the Channel Islands' ecology.

Men in ball caps and light jackets smoke on the park benches and stare in my direction. I look from the men to a nearby mural depicting *campesinos* in black clothing and woven straw hats bent over crops at dusk. I nod at the men and continue to another wall. A winged Aztec god appears to run at the viewer; his muscled brown legs with their lean tendons strain with the effort to stay ahead of the Spanish galleons and Jesuit missionaries fast approaching from the sea. A black sky looms behind the ships' white sails. A dove flies from the god's right palm, and from the left bursts white flame. I think of Juan Rodríguez Cabrillo. He could be in one of those ships.

In 1542 Cabrillo led his own expedition northward from Mexico's modern-day Manzanillo into southern California's waters. He commanded a flotilla of three ships, each occupied by Indian and black slaves, as well as sailors and soldiers, and stocked with enough provisions, including pigs and other livestock, to sustain them for two years. In less than a year, his crew made the first European contact with the Chumash, which was not a congenial meeting.

A man approaches me as I photograph the murals. He shakes my hand and introduces himself as Miguel. He wears round wire eyeglasses and a multicolored stocking cap pulled down over his forehead and graying hair. He's Chumash and Mexican, he says, shaking my hand again. He asks if I'll take a picture of him.

"Okay, sure," I say. "Where do you want me to take your picture?"

He smooths his navy blue shorts and gray sweatshirt. "Here, on this bench," he says, motioning to a bleached-out wooden seat behind him, where flecks of blue paint peel away and expose splinters.

Drumstick ice-cream wrappers, a half-smoked cigarette, and a yellow straw lift and roll in the wind around Miguel's feet, wrapped in brown leather sandals with broken-down heels. I focus the Pentax lens on his salt-and-pepper Fu Manchu mustache and soul patch.

"Okay, you ready?"

Miguel nods and asks me why I'm in Santa Barbara.

"I'm kayaking around the Channel Islands," I say.

"Man, my people are from those islands, you know," he says. "I haven't been out there yet, but I'm gonna go."

I nod to show I'm listening.

"There's living Chumash that want to participate in taking care of those islands, but the park service isn't interested. They don't want to talk to living Chumash. They'd rather just put all our people's baskets on display in some museum because it's safer to deal with the dead."

My brows furrow as I recall a recent conversation with Ann Huston, chief of cultural resource management for Channel Islands National Park. For the last five years, Huston has worked closely with many Chumash people who have helped to interpret archaeological sites, aid in reburials, lead programs for visiting groups, and provide input on park planning. Huston carefully pointed out that many Chumash factions exist, and just because Chumash share a cultural history doesn't mean they all agree on how the Channel Islands should be managed. Some groups want the islands under exclusive Chumash management and out of government hands altogether. Huston works with the only federally recognized Chumash tribe, those living on the Santa Ynés Reservation. These Chumash are verified descendants of islanders, their lineage traceable through mission records. These Chumash act as consultants regarding cultural sites and are pleased with the protection that the park service provides the Channel Islands. Regarding conflicts within the Chumash community, park service policy remains simple: Stay out of it and welcome any Chumash who would like to participate in the management of their ancestors' home.

John Anderson, a specialist in the philosophy of education and the author of several books on Chumash culture, argues that there's a dearth of cultural education offered to island visitors and that information focuses almost exclusively on the islands' ecology. While some of Anderson's assertions are in dispute, Huston does agree with his allegation that the public lacks cultural education. She works to integrate more of the islands' cultural histories into visitor education. She also agreed with me when I suggested that perhaps the park could do more to emphasize the direct relationship between cultural and ecological history on the Channel Islands and the ways that Chumash and other human inhabitants shaped their environments.

Miguel looks into the lens and smiles without opening his mouth. Only his cheek muscles lift slightly as I snap the photo. "Now I'll take a picture of you," he says.

"All right," I say, shy and unaccustomed to standing in front of the camera. I give Miguel the Pentax, and he clicks the shutter as I smile in the last light of the day.

"Send me that picture," he says, and scratches his address on a Drumstick wrapper with a dull golf pencil. We shake hands again and say goodbye.

Later, when I see the developed photo, I notice the mural over Miguel's shoulder. It depicts a butterfly with blue eyespots on its wings, markings that deter predator attacks. But the painted eyes are more humanlike, animated, as if the spots served as more than colorful warnings and the butterfly, like surviving Chumash, can watch its back for approaching danger.

Before leaving the park, I take a photo of Miguel's friends reclining on a picnic table and wearing Dodgers ball caps. Behind them, another mural depicts Jesus cradling a dying Chumash man, whose body appears to be enclosed in a dented brown garbage can labeled "Santa-Barbara Parks."

These murals are a response to a colonial legacy funded by the Spanish Empire and spearheaded by Juan Rodríguez Cabrillo. In 1937, a Portuguese organization honored Cabrillo as the "discoverer of California" and erected a monument to the explorer on San Miguel Island. The monument to Cabrillo also marks the site and date of his death, January 3, 1543, on San Miguel, or "Tuqan," as the Chumash called the island. Cabrillo's crew had camped on the island for three months, engaging in skirmishes with Chumash villagers, who did not like the Spaniards' presence on their island or the diseases and ideas of morality they brought with them. Cabrillo managed to survive several of these rows, but his luck didn't hold. In his last conflict with the Chumash, Cabrillo broke a leg, and the wound abscessed. He died from infection.

Cabrillo, still a celebrated figure, crosses my mind as I run through a clearing in four lanes of traffic on Cabrillo Boulevard. Facing the ocean, I ruminate on the legacy of exploration along this stretch of the California coast. While it seems unfair to heap responsibility for the extinction of the Chumash culture and for early nonnative plant and animal introductions on a single man, Cabrillo's voyage to the Channel Islands began a practice of introducing Old World livestock and plants to the islands, some by mistake, others intentionally.

I wander the boardwalk to the end of the pier, where vendors sell oyster shooters and fried fish. Juvenile brown pelicans sometimes laze around the wharfs of southern California and feed on the cast-off scraps of fish cleaners. I soon see one, browner than an adult bird, squatting on a bench less than an arm's length from the lap of a woman tourist whose husband takes pictures. The bird seems undisturbed, in a resting but alert position, its long tapered bill beneath its left wing.

I walk back to the hotel and book passage to Santa Cruz Island with Island Packers, the official park service concessionaire, for two persons and *The Grebe*. Tomorrow morning, we'll drive south to Ventura Harbor, where *The Islander* will depart for Scorpion Anchorage on the eastern end of Santa Cruz Island. We plan to camp there and then kayak eight miles to East Anacapa Islet.

The next morning, amid the revving of *The Islander*'s engines, deckhands balance a jumble of sea kayaks, including ours, along the stern railing. When the captain steers the boatful of passengers out of the slip, I scan the surface waters of the channel for signs of wildlife. It's not uncommon to see blue whales breaching in the Santa Barbara Channel. The clipper picks up speed, and, as if they'd been waiting, common dolphins climb the watery ladder from the channel's depths and burst through the bow waves. They surf the jade waters and occasionally rock from side to side, peering with one eye at the people pointing and smiling off the sundeck. The dolphins' sides, decorated with tan and cream-colored brushstrokes, split the blue-green water as these acrobats dive and surface. Hoots and laughter from fifty or more delighted, wave-splashed people fill the air around the boat.

After the dolphins lose interest as the boat slows, I towel off and slide into a booth where someone has left a copy of *The Santa Barbara Independent*, a local newspaper. Buried in the center of the weekly, a short piece describes *Fiesta Pequeña*, a Santa Barbara gala that opens Old Spanish Days, a celebration of Spanish influence and the work of Franciscan friars along the California coast. The article revisits fiesta events of the year before, when James Navarro, an eighteen-year-old Chumash man from nearby Santa Ynés Reservation, interrupted the musical performance of his uncle and cousins, who thumped drums traditional to the Sioux of the Midwest and danced non-Chumash style. Navarro faced the audience from the front steps of Mission Santa Barbara and denounced Old Spanish Days as a celebration of Chumash cultural disintegration. He addressed his relatives, who played not the Chumash clapstick and rattle but instruments associated with the stereotyped "Hollywood injun." The Chumash per-

formers likely played the drums because drums are most familiar to specta-
tors weaned on Hollywood representations of tomahawk-wielding Indians
in feather headdresses. Navarro accused the audience and his relatives of
condoning genocide, his words trailing off as police dragged him from the
stage, charging him with disturbing the peace.

James Navarro might say that the National Park Service has all but
forgotten that the Chumash people once lived on those islands, now fast
approaching out *The Islander*'s window. The Chumash would likely still be
there had the missionaries not forcibly removed them.

Just as the Chumash have been displaced by European exploration and
settlement, the convergence of three separate historical events has tipped
the balance in favor of exotic species on Santa Cruz, endangering nine
native plant species, and decimated the endemic island gray fox, whose
populations have fallen by 90 percent since 1993.

The story of how the island gray fox numbers plummeted from approximately 1300 to 100 individuals in less than a decade began a century ago
with the discovery of dichloro-diphenyl-trichloro-ethane (DDT) by German chemist Paul Müller. By 1939, DDT was in common use as a panacea
for all insect pests, which occupy the lowest position in many food chains.
As DDT moves up the food chain, it becomes increasingly more concen-
trated, having the most drastic effects on top predators within ecosystems.
Pioneer conservation biologist and writer Rachel Carson made the deadly
effects of DDT and other chemicals public knowledge in her book *Silent
Spring* (1962). Carson's book led to the ban of domestic DDT use in the

United States on June 14, 1972, when then administrator of the Environmental Protection Agency, William D. Ruckelshaus, signed the order.

However, even with this ban in place, the damage had been done. The bald eagle, one of two dozen nesting pairs native to Santa Cruz Island, had already begun disappearing from the Channel Islands by the 1950s and 1960s, when DDT caused their eggshells to thin and then crack before hatchlings were ready to emerge.

Southern California and its offshore regions were the largest dumping grounds for DDT. When the ban went into effect, companies like Montrose Chemical Corporation, which operated a DDT manufacturing plant in Torrance, ended up with excess DDT that they didn't know what to do with. To get rid of it without paying high disposal fees, they dumped it along with PCBs into Los Angeles County sewers, which emptied straight into the Pacific Ocean, where these chemicals continue to wreak havoc on the food chain. Montrose also disposed of hundreds of tons of DDT into the ocean near Santa Catalina Island.

DDT stays in the environment for a very long time. Using the bald eagle as our example, here's the breakdown: if a bald eagle eats a single DDT-contaminated fish, it takes approximately eight to twelve years for this raptor to metabolize *half* the DDT it ingested when it ate the fish. In addition to fish, dead pinnipeds like California sea lions make up a portion of the eagle's diet. Pinnipeds concentrate pollutants in their ample fatty tissue, and because DDT is cumulative, meaning that the body doesn't excrete it, this chemical builds up as the mammal consumes more contaminants over the course of its lifetime. When a bald eagle scavenges the carcass of a seal or sea lion, it ingests potent quantities of toxic chemicals like DDT, continuing the DDT-contamination cycle, taking years off the eagle's life, and reducing the chances of that eagle producing healthy offspring.

According to Kate Faulkner, chief natural resource manager for Channel Islands National Park, reintroducing the bald eagle to Santa Cruz Island has been part of the park's agenda for a long time. Releasing them into the environment is the easy part, she said, because bald eagles have lived on these islands for thousands of years. It's keeping them alive that poses the greatest challenge. A trustee council made up of three federal and three state agencies has allocated settlement monies from the Montrose Chemical Corporation lawsuit (the Montrose Settlements Restoration Program) to maintain pairs of bald eagles on Santa Catalina Island. So far, they haven't managed to safeguard a strong population. The problem may be that the environment can't support them, Faulkner explained. Biologists have launched a five-year feasibility study to determine whether there's still too much DDT in the environment for the eagles to self-sustain on the northern group of Channel Islands.

As the boat approaches Scorpion Anchorage, ending our one-and-a-half-hour, roughly twenty-mile crossing, California sea lions pop their heads above the ocean's frothy surface. I leave the cabin to see the doe-eyed creatures break the surface like jack-in-the-boxes and rotate their heads to get their bearings after long periods underwater.

The Islander's engines slow near the dock on the eastern end of Santa Cruz Island, and the ocean lathers. This eastern end of the island, the tip of the scorpion's tail and the 24 percent of the island owned by the National Park Service, bobs in front of me. The Nature Conservancy, a private, nonprofit conservation group, owns the larger chunk of Santa Cruz. The Nature Conservancy recently donated an additional 8500 acres to the park, and visitors are restricted to this 14,733 acres managed by the government unless granted special permission by the conservancy to access other parts of the island.

Bruce and I unload the kayak from the stern of the boat and lug it to the back of the beach. It's warm, in the upper 80s. Western gulls and American oystercatchers squall overhead as I tug my fleece sweater over my head and knot it around my waist. Kayakers in sit-on-top kayaks paddle near the shoreline. They hoot and yell, lobbing small objects onto each other's boats. Piling our gear on the beach, Bruce and I wade into the surf to find out what the paddlers are chucking. The ocean blushes the hue of Bing cherries as we near what seems like the edge of the mass. Breakers toss a burgundy-colored crustacean against my thigh, and I leap backward to peer down at a three-inch-long creature that looks like a miniature lobster. It's actually a red pelagic crab. This far north, a swarm of these crabs indicates an El Niño event, when southern waters move northward and nutrient production in the ocean declines. The giant squid of the Sea of Cortés feeds on this red crab, and I wonder if diablo rojo followed the herd of crabs north.

Hundreds of gulls flock to the swarm bobbing on the surface just off-shore. In an attempt to escape the hungry birds, the crabs splay their limbs and dig into the water with their wide, curved tails. Then, in one sweeping motion, they hug their legs to their sides and propel their bodies backward. Even so, the birds gorge on so many that they can barely fly.

Bruce and I slog out of the surf, our arms loaded with camping gear. Turning our backs on Scorpion Harbor, we trek the narrow, dusty road through historic Scorpion Ranch, a once-thriving sheep farm on the island. We pass an outbuilding with its sunken tin roof layered with tree chaff a foot deep. At one corner of the building sits a rust-red two-bottom disk plow wreathed by a tangle of grapevines escaped from the island's long-abandoned vineyards. In the 1880s, long after missionaries removed Chumash islanders from Limuw and placed them in the custody of Mission Santa Barbara, a man named Justinian Caire maintained 150 acres of vineyards here. He planted olive and eucalyptus trees, and by 1890 he'd raised and sheared more than 50,000 sheep. His retired farm implements surround us as we hike into the mouth of Scorpion Canyon. Grasses grow through the old horse-drawn disk harrows, wagons, barley seeders, and hay rakes.

The Gherini family, descendants of Justinian Caire, also raised sheep on the eastern end of Santa Cruz Island until 1984. Although the park service moved the last sheep off the island in 1990 and purchased the family's land rights, evidence of the ungulates' presence remains visible throughout the island. Along the hilly rises of Scorpion Canyon, sand pits like the traps on a golf course splotch the mountainous terrain where sheep grazing caused erosion of the landscape. More than a decade after sheep removal, these blonde craters refuse the seedlings of native vegetation and remain nude.

These cavities are just a few of the ghosts of immigrant settlers on Santa Cruz. There are many more, some of which managers plan to keep intact for the public as part of the island's cultural heritage. Later, I would understand that valued items managed and maintained as elements of the island's cultural landscape included more than old farm implements.

In Scorpion Canyon Campground, we pass crowds of campers as we chuff through eucalyptus leaves toward an open site. Dudleya, or "Santa Cruz Island live-forever" as it's called here, swells outward from canyon walls, its whorl of succulent green leaves catching my attention as I lug my gear toward the campsite. Before I can even drop my load, an island scrub jay makes an appearance. Living only on Limuw, this jay, blue as lapis lazuli and larger than its mainland cousin, is a unique species genetically distinct from the western scrub jay and the Florida scrub jay. In fact, the western and Florida species, separated by 2000 miles, are more genetically alike than the island and western varieties, separated only by a slender band of coastal waters.

The scrub jay hops nearer, curious and unafraid because it doesn't have the predators that its mainland cousins do. I make kissing sounds against the back of my hand. The bird responds by parading along its branch and then descending lower. It peers at me out of one eye. Although perched in a nonnative eucalyptus tree, this scrub jay depends on the endemic island scrub oak, nesting in and feeding on the tree's acorns, which it disperses, allowing new trees to sprout.

The scrub jay flies back up-canyon, where the scrub oaks become dense, and Bruce and I drop our gear beneath a eucalyptus grove and set up the tent. I glance around at the colorful tarps, the kids and coolers, and the clotheslines strung between low-growing eucalyptus branches. As a child on vacation in California, I never thought about whether eucalyptus trees belonged here. They've been here as long as I've been alive, and during the lifetime of my mother and grandmother. In fact, "euks," as they are often called, arrived from Australia in the middle of the nineteenth century, during the girlhood of my great-great-grandmother. People deemed them "wonder trees" because they grow from seedlings to forty-foot-tall trees in just three years. Their history has made them, in the eyes of many Californians, heritage trees deserving of strict protection. In some areas of the state, if a person fells a healthy eucalyptus tree on his own property, a deed often reported by a watchful neighbor, the owner is assessed a fine of $500. It's in the law books.

I understand the protective stance toward these paternal softwoods, many of which settlers planted in 1850, the year California established statehood. On Santa Cruz Island, the eucalyptus trees are at least 150 years old, planted when Chumash still lived and spoke their native language here. But the trees don't belong here. Some ecologists, referring to the trees as biological "vacuums" because they don't provide animals with

The Scorpion Canyon Campground supports a grove of heritage
eucalyptus trees.

food, have documented a 70 percent decline in bird diversity where eucalyptus trees dominate. Far from coexisting peacefully with native plants,
eucalyptus trees secrete toxins preventing native vegetation from growing
in their vicinity. Firefighters call them "gasoline trees" because the species
thrives on fire. Its flammable oils encourage the trees to burn and disperse
seeds. Blue gum, the most common eucalyptus growing in the United
States, often explodes when on fire, scattering not only seeds but flam-
ing branches in all directions. Naturalist and essayist Ted Williams noted,

"Living next to one of these trees is like living next to a fireworks factory staffed by chain-smokers," and he quoted the famous photographer Ansel Adams, who once scattered a group of Boy Scouts planting eucalyptus seedlings, announcing "I cannot think of a more tasteless undertaking than to plant trees in a naturally treeless area."

But while the eucalyptus poses a significant threat to healthy ecological function, this tree also provides habitat for monarch butterflies whose California migration path might have been eliminated due to human population growth and felling of native trees for housing developments. This is one of those gray areas rarely acknowledged in invasive species management, a difficult case in which complete eradication of the offender isn't necessarily the best course of action. The eucalyptus tree, along with a handful of other invasive species, harms certain ecological functions while enhancing others. In the tree's defense, I have to admit a guilty appreciation for the shade provided by these behemoths. Without them, we campers would sear in the August sun.

In the end, the National Park Service has the final word on the fate of the eucalyptus trees on the Channel Islands. When I spoke with Kate Faulkner, she explained that along with the old ranch and its farm implements, the eucalyptus trees will remain as part of cultural landscape maintenance. "We'll pull up any seedlings that take root," she said, "but as to the older trees, we've decided we can live with them." Ted Williams took a different stance, "These trees are beautiful and precious. In Australia."

Following a dinner of shelled peanuts, smoked salmon, and saltines,

Bruce and I walk to the landing dock in the dark to peer into the black water. Below the platform, a riot of fish bolts in one direction and then another. These quick movements create milky white bursts of light, one flash followed by another and another. The phosphorescence is caused by microscopic single-celled algae (dinoflagellates), which give off light when disturbed. The fish charge through the water, igniting it like cloud-to-cloud lightning beneath the ocean surface.

The next morning I wake to the crackle of dry eucalyptus leaves outside the tent. "Fox?" I think, and my breath quickens. But I know better. The island gray fox, while a common sighting for visitors at one time, is rarely seen now. Its population has plummeted in recent years. On San Miguel Island, for example, fox numbers dropped from 450 in 1994 to approximately fifteen animals in 1999—435 animals disappeared in five years. By 2002, only one fox remained on San Miguel. On Santa Cruz, a much larger island, the drop in island gray fox numbers is also staggering. To preserve this creature, island managers had to learn the source of the problem, and they've only recently solved that mystery. The disappearance of the fox resulted from a series of events that began with Spanish colonization along the California coast and continued with island settlement by westward-moving American families. No single thread of this story is sufficient by itself to threaten the island gray fox. Even two of these threads when combined couldn't push the fox to the brink of extinction. Only all three, when placed together, tell the whole story.

The use and then the offshore dumping of DDT, responsible for the decline in numbers of the bald eagle, is also implicated in the plight of the fox. Biologists discovered that the near vanishing of Knt-y, the Chumash name for the world's smallest fox, is an unforeseen consequence of bald eagle decline and "squatting" by the nonnative golden eagle that moved in during the 1990s. Because bald eagles and golden eagles rarely share territories, when the bald eagle vanished, the golden eagle was able to populate the island. And while bald eagles do not prey on foxes—they eat only fish and carrion—golden eagles do.

What first enticed North America's largest predatory bird to take up residence on Santa Cruz Island? The answer has everything to do with urban sprawl and the animal rustling outside my tent. Unzipping the fly, I poke my head out and see the backside of a piglet, its hide mottled with gray and black splotches. Sniffing at coolers and tent stakes, the piglet stops abruptly at the pant legs of the man occupying the tent site next to ours. The man stands still, gazing down on the coarse hair of the animal. The piglet halts his snuffling and looks scared, as if he senses he's not where he should be. And indeed, he isn't.

In the 1850s, if not earlier, white settlers introduced the pig—a mix of the European wild boar and domestic breeds—for husbandry and hunting on Santa Cruz Island. The pigs developed a feral population that exploded, rooting up native plants and destroying Chumash archaeological sites.

"You lost, little piggy?" asks the man, and the piglet runs into the brush, then turns around to listen.

I climb out of my bag and slip on shorts and shoes. "Cute, isn't he?" I say

An invasive golden eagle
captured for relocation

to the man, and the piglet backs into the brush farther still at the sound of
my voice. "Pigs aren't supposed to live on these islands, but they are cute."

"What isn't cute when it's a baby?" the man says. "You see its mama?"
Not waiting to hear my answer, the man says, "Now she's something ugly.
Big as a shack, and real mean. She's over on that ridge there, somewhere in
those prickly star thistles."

I lift my eyes to the hillside opposite the campground, wishing I could see a soaring bald eagle. But aside from a patch of invasive yellow star this-tles silhouetted against the sky, I see nothing but holes and patchy areas of dirt where pigs have foraged through the brush, pulling everything out by the roots. Despite the destruction the feral pigs have caused, pigs and foxes can live together without endangering the fox. It's the third thread of this story, the one that originally sent the golden eagle packing to the Channel Islands, that finally spelled doom for the island gray fox.

Although federally protected by a 1962 amendment to the Bald Eagle Protection Act of 1940, the golden eagle has been squeezed out of its native home by urban sprawl in Los Angeles and surrounding areas. But, as further amended in 1978, the act "authorizes the Secretary of the Inte-rior to permit the taking of golden eagle nests that interfere with resource development or recovery operations." Thus, the only range left in southern California for this bird of prey is 72,000 acres of land off-limits to devel-opment. This may sound like a lot of land, but it is fragmented habitat, a spotty few thousand acres here and a few thousand there. Mated pairs of golden eagles in the western United States require an average territory of 8000 acres. Thus, 72,000 acres can support only nine nesting pairs, a very small population. A healthy population might contain upward of ninety golden eagles, mostly adults, with perhaps thirty or so of those ninety juvenile birds.

Golden eagles can never go home again, and the protected Channel Islands must have appeared like an oasis. Some people argue that because golden eagles solved the problem of habitat loss on their own by moving

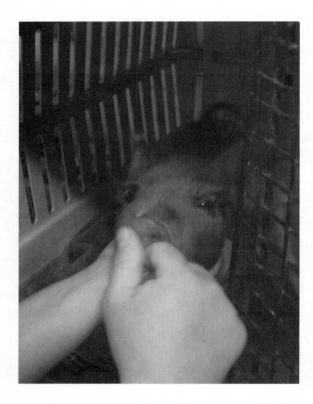

An invasive feral pig. It's hard to believe a cute piglet could do so much damage.

onto the Channel Islands, they should be considered native and left alone. But others, like Kate Faulkner, maintain that although humans didn't hand-deliver the golden eagle to Santa Cruz Island, without human population pressures and the degradation of golden eagle habitat, these raptors would never have left the Los Angeles area. "If the bald eagle population hadn't been decimated by DDT, golden eagles would never have come here in

Feral pig
rooting
damage

the first place," she told me. "Without the pigs, the golden eagles probably wouldn't have stayed to discover that the island fox is easier pickings."

Faulkner explained that the golden eagles began making day trips across the channel to prey on the feral pigs. Then the aerial hunter found the island gray fox, no larger than a house cat and weighing in at less than five pounds, an easier target than the pigs. Unlike its mainland cousin the gray fox, which hunts at night to avoid predators, island foxes have had no predators to fear and have evolved a daytime hunting habit. The fox didn't know to look skyward for potential marauders.

I shuffle around in the soil for the best light, training the camera lens on the disoriented piglet, when I feel my sneakers sink into something soft, like mud. It's my first mound of pig apples, golf-ball-sized and brownish green. My fellow camper laughs as I scrape the dung from my soles. He tosses dried ramen noodles on the ground. "That should keep him around," he says, nodding to my camera. "You know, they ought to have hunts out here. I've been coming to this island for years now. Used to be able to hunt these four-legged rototillers with a bow."

Sport hunting, according to Faulkner, doesn't work because sport hunters pick off the easiest targets—pigs along the roadsides, not in the interior, where the majority of the pigs live. Every last pig has to go. To rid the island of the pigs, the National Park Service and The Nature Conservancy assert that they have considered alternatives to a large-scale hunting campaign, from moving the pigs off the island to contraceptive measures to constructing an enclosure on the island and deeming it a "pig sanctuary." During a public presentation about the pig eradication project, the new park superintendent, Russell Galipeau, addressed the idea of sterilization and contraception for the pigs. "It's not an option," he said. "There is currently no contraceptive that can be shot into a pig from a distance, and there's no method for marking individuals who have been sterilized. We would have to handle every individual pig in order to deliver a contraceptive. It's just not feasible. Fast-reproducing animals like pigs are simply not candidates for contraception." After considering the alternatives, these organizations rejected them as impractical and ultimately contradictory to their goal of island restoration.

This is where restoration gets tricky. The park service opted to keep some eucalyptus trees for heritage reasons, but chose not to keep any heritage pigs. In other words, one element of the island's agricultural and husbandry history remains, while the other must go. Sound hypocritical? Not really, if we consider that the feral pigs cause greater damage over a larger area of the island. While park rangers pluck any stray eucalyptus seedlings and maintain only a few trees, controlling pig reproduction and environmental destruction is nearly impossible.

"Four-legged rototiller" refers to the pig's tendency to dig two-foot-deep craters while looking for the bulbs of nonnative sweet fennel, intentionally introduced to the island for its essential oils. "Wherever the pigs go, sweet fennel follows," said Galipeau. "There's so much fennel on the island, you'd think we'd established a plantation out there." Feral pigs don't just endanger the island fox, although this is the angle presented to the public because people are drawn to furry, cuddly-looking creatures. The fact is that the pig also threatens nine native plant species. Of these, only four *individual* plants of the endemic island barberry remain. The pigs uproot rare plants ranging from the endemic island oak to experimental plots of Santa Cruz Island live-forever and Hoffmann's rock cress, reintroduced by Dieter Wilken, a horticulturalist I spoke to. Wilken told me that the only way to remove the pigs is to kill them. "Pigs aren't easily caught in traps, and they carry diseases that preclude their transport to the mainland in any kind of 'rescue' effort." The pigs could carry pseudo-rabies, and the state won't grant permits for pig relocation to the mainland.

A patch of invasive
sweet fennel

"The park service would sooner shoot these hogs from a helicopter," sniffs my camping neighbor with disgust. "Or they'll poison the island's water supply rather than involve the public and let a few skilled hunters take home some pork to feed their families."

I excuse myself and creep through the brush to a chain-link rectangular holding pen. On the front of the corral measuring nine by six by five feet, a laminated sign reads, "FERAL ANIMALS. KEEP BACK!" The trap works by luring a pig through the in-swinging door with bait; once the animal is inside, the door clicks shut. The enclosure is empty now. Long braids of frayed and sun-bleached rope drape the entrance. The park service has a different plan for the pigs.

My camp neighbor is partially right. The park service will eradicate the island's more than 2000 feral pigs in one sweep, erecting forty-five miles of pig fencing to confine the island's population into six well-contained zones. Norm MacDonald and his associates from ProHunt New Zealand won the bid for the job. There were fourteen bidders, mostly U.S. companies, but ProHunt was the cheapest and was willing to sign a contract to kill the swine to the very last individual. The park service is putting up $2.6 million and The Nature Conservancy $2.4 million to complete the eradication. ProHunt New Zealand, over a period of twenty-seven months, will clear the island of the pigs zone by zone, shooting and literally hounding them until they're all dead.

Not only will the public not be able to hunt the pigs, but the park service can't even afford to donate the pork to soup kitchens to feed the homeless. The pork would have to undergo expensive screening procedures because the island hogs do have diseases, as Dieter Wilken pointed out. But according to Kate Faulkner, disease is the least of the problems. FDA regulations for handling the meat are so strict that it's just cost-prohibitive to give out or even sell the pork. "We'd be better off taking that money and buying the homeless a million dollars' worth of pork chops at Von's Supermarket," she said.

There's an irony here—one generation of hunters and farmers brought the pigs to Santa Cruz, while another generation gets hired to kill them off. In an article for the *Santa Barbara News-Press*, former superintendent of Channel Islands National Park, Tim Setnicka, asserted that for the first time in its history the National Park Service has been forced to rethink its secretiveness in eradicating nonnative animals. Management activities at previously little-known parks like the Channel Islands, formerly "hidden offshore from public view," are no longer concealed from public scrutiny because tourism has increased there exponentially. The Santa Cruz Island pig project is the epitome of every exotic-species eradication program in the world. But eradication is ugly. In fact, with animals there are few "clean" kills. Setnicka wrote, "We frequently gut shot and wounded pigs who escaped. When sows were shot, their piglets were caught by dogs or we chased them down on foot. The dogs [wearing Kevlar vests] . . . would tear into and mangle the smaller pigs. The larger pigs would fight the dogs, occasionally injuring or killing one. Due to the close quarters, pigs were caught by their hind legs and then were knifed or beaten to death."

The primary reason for shielding the public seems to stem from a well-justified fear that animal-rights groups will attempt to halt eradication efforts. Setnicka has argued for an advisory board, a liaison between the National Park Service, other governmental organizations charged with ecosystem management, and the general public, which, he pointed out, shares in the park's ownership. Such a liaison could garner more support for actions that need to be taken to restore native ecosystems, but this kind of committee will work only if the public is fully informed about the science behind conservation decisions, as well as the messy details of eradication. Setnicka continued, "Watching an animal bleed to death after sticking a knife in their jugular vein is a horrendous sight. You watch the life drain out of their eye, which becomes dull as they die. This is an impossible image to sell to the public or politicians, which is why no photos are allowed."

Shouldn't the general public know the ugliness of the process, the suffering that an individual animal, introduced to a foreign ecosystem, goes through? Awareness of the eradication process might drive home the need to prevent species introductions in the first place. According to Setnicka, putting an end to clandestine eradication activities by establishing a liaison between government organizations like the park service and the general public would not doom the pig-removal project, as feared, but instead ensure its success by educating concerned citizens about what's at stake for native and exotic inhabitants of a place. Such an advisory board might even be a model for eradication projects around the world.

Setnicka has certainly made a case for involving the public in restoration decisions, rather than pushing large, costly, and divisive projects on them. However, without a measure of literacy about bioinvasion, without an understanding of what an invasive species can do to a native ecosystem, the chasm between animal-rights groups, the general public, and government scientists and land managers will redouble. The field of invasion ecology will remain confined to a few specialists in the know, working behind the scenes to eradicate unwanted species from parks and other areas slated for restoration.

Getting rid of the pigs is a big move toward saving the island fox, but it's just one part of a four-part program that also includes continuing the captive-breeding program for the island fox, reintroducing bald eagles, and relocating the golden eagles to the eastern side of the Sierra Nevada, a habitat already inhabited by this species.

Relocating the golden eagles has been partially successful. Satellite telemetry has shown that none of the released goldens have attempted to cross the Sierra Nevada and return to the Channel Islands. Some of the golden eagles suffered shock from the transition and died, however, prompting accusations of senseless killing. Yet, the removal of thirty-five golden eagles from Santa Cruz Island in 2003 and six more in 2004, using live-trapping methods, has allowed the fox population to begin its slow recovery. In October and November of 2004, twenty-three captive-bred island foxes were released into the wild, with the eventual goal of re-establishing a self-sustaining wild population.

When I return to the campsite, Bruce is already gone. I shuffle down the sandy road to the beach, where I spot him snorkeling just offshore. A California harbor seal, with its shimmering brownish black skin, swims beside him. Bruce sees me gawking and points a finger at the seal circling him. He indicates that the seal has touched its nose to his goggles. These seals have had a lot of contact with people, I think, and I'm not so sure that's good for them.

The endangered Channel Island fox is preyed upon by the invasive golden eagle.

This morning Bruce and I plan to paddle roughly eight nautical miles from Scorpion Anchorage on Santa Cruz Island to Arch Rock, at the eastern tip of Anacapa Island. We want to see Anacapa, the site of recent black rat eradications. We'll cross the Anacapa Passage and parallel the entrance of each of the island's many sea caves before reaching Landing Cove. East Anacapa, the only islet where visitors are allowed, also represents a major shift in island management, from an emphasis on large-scale ranching to an age of habitat restoration. Along with this transition comes antagonism and death.

Bruce and I lug the boat down to the shoreline, the soles of our neo-prene shoes sliding on crushed stone and shell-grit. We climb into our cockpits and push off, paddling out of Scorpion Anchorage through a clus-ter of red pelagic crabs. "Sea locusts!" Bruce shouts.

As we round the eastern tip of Santa Cruz Island, paddling toward Cathedral Cove and Anacapa's sea caves, we glide over shallow reefs where bright orange Garibaldi fish guard their territories and red sea urchins lie still, suctioned to their undersea boulders. Kelp fronds rise up from their dark origins toward the sunlight.

Rat Rock on West Anacapa comes into view, and I stop paddling for a moment to look through binoculars at the landmark. I recall Kate Faulkner telling me that island rehabilitation means different things to different people. The objective might be the same, but people disagree about how to get there. The primary objective of the park service's Anacapa Island Restoration Project is to eradicate the nonnative black rat and other exot-ics to restore "native balance" on the island, as defined by species that lived on the islands prior to European colonization. That excludes the black rat.

Rattus rattus, the most common nonnative rat to establish island popu-lations throughout the world, adopted Anacapa as its home. Black rats are known predators of island seabirds, eating eggs and young. Even the natu-ralist Charles Darwin, lover of all things English, bemoaned the introduc-tion of rats to islands because they tend to kill off native wildlife. While in New Zealand, Darwin noted the numbers of exotic species replacing

An invasive black rat pokes its head through crumpled newspaper.

indigenous ones. He could have been writing about Anacapa Island: "In many places I noticed several sorts of weeds, which like the rats, I was forced to own as countrymen."

For more than twenty years, Harry Carter has studied the Xantus's murrelet, a small, slender seabird with a black back and white cheeks, throat, and underbelly. I had spoken with him at the Pacific Seabird Conference in Santa Barbara. He explained that the black rat preys heavily on the eggs, chicks, and even the adults of this small seabird. Although the number of murrelet nests on Anacapa Island ranges from 200 to 400, Carter remained certain that "the population of murrelets will likely be ten times the current size when rats are gone."

The black rat preys on more than the Xantus's murrelet. Western gulls, the endemic Anacapa deer mouse, lizard, and salamander populations have also dropped as a result of rat predation. Island managers like Kate Faulkner assert that black rat removal will prevent these species from going extinct.

"Who could love a rat?" wrote Tim Setnicka. "Well, as it turned out, lots of people." The park service didn't figure on The Fund for Animals, a national animal-rights organization that denounced eradication plans for the black rat as a crazed "religious fervor" to get rid of all exotic species. I had chatted with Michael Markarian, the organization's president, a few weeks before this trip. Friendly and eager to discuss the issue, he fielded all of my questions with aplomb. "Our main concern is that rats are living animals who can feel pain and can suffer, and they should be treated humanely whether they are considered 'integral' or not."

Native to the Asian continent, the black rat likely arrived in Europe during the Crusades. Evidence suggests that the true origin of the plague that killed half of China's population by 1393 was a type of flea living on the black rat. Crusader-soldiers brought the plague from central Asia to Europe. As they fell with fever, their bodies covered in black pustules, their fellow warriors catapulted their blackened and disease-infested corpses over the town walls of cities they assaulted. The epidemic moved from village to town, spreading across Europe and eventually killing 200 million people. Although there is disagreement about the origins of the Black Death, people still associate rats with sickness and all things foul. As Bruce and I paddle on in silence, I can't help thinking of Robert Browning's "The Pied Piper of Hamelin," often read to me as a child and likely based on a German legend passed down from the Crusades:

Rats!
They fought the dogs and killed the cats,
And bit the babies in the cradles,
And ate the cheeses out of the vats.

Rats are vermin in the minds of many people. They are objectionable creatures that follow human societies wherever we go. Also commonly known as the "ship rat," the black rat followed Spanish conquistadors to North America aboard vessels of the Spanish Empire. It continues to cohabit with humans in attics and barns, warehouses, and even wall insulation, largely undetected.

Inside the Anacapa Passage, Bruce notices a sea lion pursuing the kayak. I swivel my head to see a set of whiskers to rival the handlebar mustache of a Hollywood lawman. The sea lion follows in our wake, a short distance from the rudder. When it tires of us, it slips in its sleek black suit beneath the ocean's surface. Marine mammals appear and leave so quickly, I often wonder if I've seen them at all.

Fitting its Chumash name for Anacapa, Anyapakh (meaning "mirage"), the island appears and disappears behind a thin veil of fog. But the mirage could just as easily refer to the island's so-called native balance. Perhaps the native balance exists, but maybe it doesn't. Even with the rats gone, people continue trafficking to and from the island, particularly the park service's researchers, rangers, and maintenance workers with equipment inside of which exotic plants and animals may be transported.

The black rat didn't travel to the island inside maintenance equipment, however, but apparently arrived on Anacapa by shipwreck. This rodent's 150-year tenure on the island may have begun with the end of the steamship the *Winfield Scott*. During the California Gold Rush, while overland travelers trudged the slow route across the continent, passengers aboard the *Winfield Scott* made the voyage in a record forty-nine days. The ship traveled from the East Coast of North America, rounded the tip of South America at Cape Horn, where Darwin had spent Christmas Day twenty years earlier, and plotted a course for California.

Darwin's disquiet about rats spreading to island communities might have been an omen. On December 1, 1853, the *Winfield Scott*, its gold seekers having disembarked in San Francisco, turned around in the bay and sailed for Panama. To save time, the captain navigated his ship, with its 300 passengers and gold bullion aboard, through the Santa Barbara Channel, rather than skirting the island chain from the outside, in open water.

Unpredictable weather. made his choice a gamble. A fog emerged around eleven at night, a thick mantle shrouding the islands. The *Winfield Scott* plowed into Middle Anacapa, wrecking off the northern side of the island. The cracked ship not only spilled its human passengers, who

Buckwheat growing along the shore of Anacapa Island

rowed dories or swam to safety on Anacapa's shores, but also hitchhiking rats, labeled "roof rats" because their claws are shaped for climbing verticals.

Because the black rat wasn't documented on the island until the

1920s, another explanation for how rats arrived to Anacapa is plausible. Kate Faulkner asserts that rats may have come over with lighthouse keepers in the 1920s, when supplies were brought back and forth from the mainland.

One thing is sure, however. Black rats are tough. They have been seen treading water two weeks after a ship has capsized, and they can squeeze through a hole the size of a quarter. Rat teeth are stronger than steel and capable of chewing through concrete, providing fuel for horror films about them. Black rats spread like an oil slick on islands. They reproduce at just three months of age, and following a short three-week gestation period, females deliver litters of up to eight young, who repeat the cycle in another three months. A stowaway pair of breeding rats, once they scuttle ashore, can colonize an island with 5000 individuals in a single year. They're adaptable to almost any set of environmental conditions, including the dearth of fresh water on Anacapa Island.

We paddle along the north side of Anacapa, where we hope to see the remains of the shipwrecked *Winfield Scott*. I picture rats swimming in the dark water, surrounded by fog. Arch Rock appears in the distance, off the eastern end of the island, looming on the horizon like a stout tree stump with a hole drilled through the middle. No shipwreck parts emerge from the surf by the time we reach the crags of East Anacapa, but the landing cove is a roiling cauldron. The only craft in the cove is ours, and we're tiny inside this fishbowl, where waves unfurl their watery carpets, lifting and dropping the kayak as they rumble past. Foam spatters the kelp beds as I strain my eyes upward, through shafts of sunlight, in the direction of the lighthouse, not built until after the wreck of the *Winfield Scott*. The drone of the foghorn splits the air inside the cove.

Young western gulls tumble from the cliffs above our heads and find their flight wings for the first time. We float toward a set of stilts supporting a landing dock ten feet above the water level. We paddle to stay parallel, waves washing over the stern of the kayak as we sidle up to the bottoms of two wooden ladders spaced evenly apart. Bruce ties the bowline to the rung of one ladder, and we unload the kayak's three hatches, hauling out sleeping bags, tent, and food sacks of tinned sardines, granola, and spaghetti. Then we haul out our nylon water bags. We hope we've brought enough. Not even the Chumash attempted to establish a village on Anacapa because it lacks fresh water.

When the kayak floats light above the kelp beds, we hoist its bulk onto the dock, using a crane provided by the park service. Bruce cranks up the bow while I use another rope to haul up the stern. We leave the kayak and paddles drying on the dock and sling bags over our shoulders for the hike up 153 steps to the top of the cliffs. Someone had warned me about these stairs, but nothing could prepare me for what I smelled at the top.

"Thousands of birds live here," I say to Bruce. "You smell them?"

"I've never wished for a head cold more than now," he says.

I let the gear slip off my shoulders and drop to one knee to inspect a pyramid of dusty bones at my feet. There's an intact wishbone and a leg bone with strings of meat still clinging to cartilage. Chicken bones.

This young western gull is part of Anacapa Island's vibrant nesting seabird population.

Gulls are scavengers, and they're efficient. Adult western gulls dive from the island cliffs and wing their way to the mainland for garbage raids. They gorge on what they can and ferry scraps to their broods. Gull chicks in clutches of two or three worry their parents with high-pitched mewls; their necks crane forward, a gesture of pleading, until the parent gull acquiesces and regurgitates a meal.

I leave the bones where the gull deposited them and lug our gear a few more paces to the white stucco visitor center, with its red-tile roof. Pelicans and gulls have coated the building's patio, benches, and picnic tables with splatters of grayish white excrement that smells like cat piss warmed over a car engine.

"This place is depressing," Bruce says. "Sure as shit, I'm not coming back here." In the island's guest book, other visitors have written similar comments. I thumb through the pages, and instead of questions about native plants and animals living on the island, I encounter entries like "You should rename this place 'Shit Island,'" and "Can't you guys clean up the poop once in a while?"

Casting a glance around the island, it's hard to imagine where the park staff would even start. Besides, this island is a breeding ground for seabirds and a park only for people who can bear to watch them while plugging their noses. More important, the National Park Service has more than enough to do in managing the balance of native and exotic species on this island without worrying about the quantity of seabird excretions. We sign our names and continue a half mile down the trail.

We pass mounds of Santa Cruz Island buckwheat, which also grows here on Anacapa, and the sticky yellow heads of gumweed. Gull chicks, with their downy polka-dot crowns, lumber around in a forest of giant coreopsis, which is also called "tree sunflower" for its yellow ray flowers that spring from trunks thick as Sumo wrestlers' thighs. We stop to watch a parent gull perch on a sign announcing "Pinniped Point" to escape the pleading fits of its two chicks. But the fledglings repeat their plaintive cries until the parent consents to leave its perch and feed them. I turn from them and face north, toward the down-sloping cliff where native plants grow on waterless and exposed rock. Just when I get excited about the numbers of native species, foxtail barley jabs me in the thigh. A settler on Santa Cruz Island first recorded foxtail on the northern Channel Islands in 1888.

In my field journal I document the presence of foxtail and carpets of iceplant species, crystal iceplant and the Hottentot fig, that smother native vegetation and spread unhindered along Anacapa's cliffs. Even Kate Faulkner acknowledged, "Iceplant is out of control. We've had volunteers to spot-treat Anacapa, but we really can't devote personnel and money to eradicate iceplant species right now. We have enough on our plate."

When we reach the campground, Bruce and I have our choice of campsites. We're alone on the island. The wind snakes along the sea cliff. We scout each campsite, and a quick look tells us that no matter which site we choose, there is no square inch free of dried bird feces. We settle on a site and find in the grass behind the picnic table a bait box for rats with "C3" hand-painted on the front. Browning's words about the rat-catcher run through my head:

Into the street the Piper stept,
Smiling first a little smile,
As if he knew what magic slept
In his quiet pipe the while.

If I'd camped along Anacapa's cliffs two years earlier, I might have felt the soft pads of a black rat's claws as it scampered across my sleeping bag, just as I felt the moist pads of a kangaroo rat's feet press my cheeks as it went about its nocturnal activity on Ventana Island in the Sea of Cortés.

Although rats have been vilified as vermin, pests, and baby killers, I don't dislike them. In fact, at one time I looked into obtaining a rat companion, assuming that they traveled more easily than dogs, particularly aboard airlines. I was wrong. Although rats don't take much space, don't bark, don't stink or defecate on the floor, airline regulations mandate that all "vermin" must travel in the cargo hold, never mind that a passenger may carry a small monkey on board, as long as the monkey's cage size allows it to fit beneath the seat in front of him.

I climb the hill above the campground and walk the trail to Inspiration Point, where I can survey Middle and West Anacapa Islets, five-sixths the total area of Anacapa Island and recently rat-free. Peering over the cliff, I train my binoculars on the spine of land between the middle and east islets. Faulkner told me that the park service took a risk in the fall of 2001, the first year of rat eradication, by killing rats only on East Anacapa as a trial effort. The park service wanted to find out if eradication could work. "We took the chance," she said, "because East Anacapa is only one-sixth of the whole island. By not treating the entire island at once, we had a chance to refine our methods."

The short window of opportunity to eliminate rats on Anacapa occurred during November and December, months when seabirds finish annual nesting cycles. Because the rats no longer had access to an easy food source, managers theorized that they would readily consume the poison bait pellets. Although the park service could work out any kinks following this experiment, splitting up the eradication process held a two-fold risk. The weather requirements for a successful eradication shrink the two-month window. Also, managers had to drop the poison bait during a five- to seven-day period of no rainfall because rain causes the bait pellets to break down prematurely.

The park service finally got an opening for their test treatment on East Anacapa in December 2001, their last chance for the year, and they went for it. A helicopter approached the island and aerially sprayed BB-sized pellets of brodifacoum, an anticoagulant rat poison containing Dekon, over every inch of rat territory. The old-fashioned method of employing bait traps like the black box in my campsite has never worked, according to island managers. Anacapa's severe topography, its cliffs where rats often nest in nooks and crannies, makes bait traps and hand broadcasting ineffective.

The National Park Service, in partnership with the Island Conservation and Ecology Group, followed the lead of New Zealand, the only other country to use aerial broadcasting to poison rats. According to Faulkner, using the poison "created habitat where native species can thrive." Eradicating rats would allow endemic deer mice, lizards, salamanders, and nesting seabirds to recoup their former numbers. Some of them had barely hung on in the company of rats. According to Faulkner, aerial spraying gets results in six days to two weeks and minimizes the deaths of nontarget species.

To protect nontarget animals such as the endemic deer mouse, which would also eat the poison pellets due to the scarcity of other foods, park service biologists consulted researchers at the University of Illinois, Chicago, to determine how many deer mice they should place in protective captivity during the poisoning to maintain a healthy genetic diversity in the population. Careful DNA analysis put the minimum number at 300 to maintain 99 percent of the genetic diversity in the deer mouse population. Not wanting to gamble with evolutionary forces any more than necessary, the park service tripled that number and trapped and transported 900 deer mice to a holding facility on East Anacapa for release following rat elimination. Any uncaptured deer mice would likely die with the rats.

I stare from the scarp onto the isthmus. This lean backbone of land is all that separates rat-free East Anacapa—now harboring double the number of side-blotched lizards and Pacific slender salamanders—from those starved rats still populating Middle and West Anacapa Islets. I imagine the black rats pacing this fifty-acre buffer zone, trying to figure out how to get to the food source. If they had, the rat's rapid reproductive rate would have allowed it to recolonize East Anacapa with ease. But the park service knew this might happen. They placed bait stations along the shoreline and the spine between the islets, as well as paraffin chew blocks to register tooth impressions. They even laid tracking board to detect rat footprints. Finally, monitors put out conventional bait boxes. "We found prints in the intertidal area, and we caught three rats," Kate Faulkner said.

Great rats, small rats, lean rats, brawny rats,
Brown rats, black rats, grey rats, tawny rats . . .
Until they came to the river Weser
Wherein all plunged and perished!

Although the deaths in phase 1 included about 1500 endemic deer mice on East Anacapa, along with at least forty-nine individual seabirds from eleven different species, the park service deemed these deaths acceptable losses. Faulkner didn't skip a beat as she explained that the objective of keeping native species viable takes precedence over individual animal deaths. "Most of those mice would have died over the winter anyway. We're talking about a short-lived species. And we also know that rats prey heavily on the mice." While monitoring for eradication success on East Anacapa, the park service prepared for phase 2 on Middle and West Anacapa Islets scheduled for November and December 2002.

The bawl of gulls in the air above my head and the shrill pleading of clownish gull chicks is the normal business of birds. But the clamor that has surrounded this island is not the commotion of seabirds, but the wrangle over eradicating the black rat. The clash is deafening. Although success in phase 1 of the eradication pleased many, The Fund for Animals and CHIAPA objected. They condemned the eradication of individual rats, mice, and seabirds as senseless slaughter. These two animal-rights organizations filed a lawsuit against the National Park Service, which they later dropped, yet not everyone conceded that phase 2 of rat extermination was inevitable or even necessary.

When I spoke to Michael Markarian, he argued that the anticoagulant brodifacoum, the poison used in the aerial spraying of East Anacapa, functions like DDT in the environment. It has a long half-life like DDT, he said, and it results in more wildlife deaths than can be documented. That argument was countered by Kate Faulkner's assertion that the park service had a custom formula of brodifacoum developed for treating Anacapa. It contained half the amount of Dekon in the regular formula. "In fact," she added, "the poison bait we used contains half the amount of rodenticide found in many rodent-control products that homeowners can purchase in their local feed and hardware stores." She also explained that the pellets broke down once aerial spraying finished. "We wanted the rats to pick them up, but we didn't want the poison hanging around," she said. The regular island fog could break down the unwaxed pellets, and in approximately two weeks the pellets went through a process of microbial degradation, growing mold as they began to bind with soil. After three or four weeks, she claimed, the pellets eroded into two nontoxic components, carbon dioxide and water.

The Fund for Animals and CHIAPA didn't buy it. Poison is poison. For them, valuing the life of the individual animal has priority over species considerations. This is where the National Park Service and these two animal-rights groups diverge sharply. The mission of the park service, to maintain indigenous life and preserve whole ecological communities, sometimes requires the sacrifice of individual animals. Markarian, in agreement that nonnative animals shouldn't be released into environments, accidentally or intentionally, nevertheless wanted the rats captured and relocated to another island or mainland location. He didn't have a specific location in mind, just that the park service should relocate them somewhere. "Black rats have been part of the Anacapa ecosystem for a century and a half," he said, "and whether they are native or not, they do not deserve to die slowly over three to ten days from a poison that causes internal bleeding." He also suggested that the park service should try using exclusion devices such as protective wiring over bird nests to guard eggs from predators.

Other voices accompanied Markarian's to decry the elimination of rats. Rob Puddicombe, a former commercial diver and now a Santa Barbara bus driver and founder of CHIAPA, said, "To me, the idea of species is just an abstract concept. Species go extinct all the time. Animals are individuals with their own personalities, not little robot machines."

Huddled at Inspiration Point, I picture the drama that unfolded in the fall of 2002. To Rob Puddicombe, an ecosystem, an integrated community of native plants and animals, remains a vague idea used by the park service to justify killing thousands of innocent animals. To him, the park service killed in the name of some abstraction, while he performed the real work of saving lives. CHIAPA's mission statement includes the assertion, "CHIAPA values the lives of individuals far above cold, categorical statistics." To prevent the carnage, Puddicombe and a companion boated across the twelve-mile sweep of channel waters separating Anacapa from the mainland.

The park service and the Island Conservation and Ecology Group had just laced the two islets with brodifacoum when Puddicombe and Robert Crawford readied their mission. When they reached the island, Puddicombe climbed onto Anacapa's rocks, carting a five-pound sack of kibble fortified with vitamin K, the only substance proven to neutralize the anticoagulant properties of brodifacoum. Puddicombe spread the kibble, clambered back into the boat, and the two men prepared to depart. Their engine stalled, leaving the men stranded, while a suspicious park ranger fast approached.

Puddicombe, placed under arrest for "feeding wildlife" and "interfering with a federal function," stood on the beach that day and read aloud a note he'd written to himself, "The National Park Service is playing God on the Channel Islands, and it lacks the wisdom and compassion to do the job."

As I walk back through the campground toward Pinniped Point, I think of something Puddicombe said during his trial, "I can just imagine those rats bleeding to death in the rain." Puddicombe's comment about playing God makes me think about our human tendency to change the natural world to suit us. Some people argue that adjudication of nature is an arrogant stance for humans to take. Who says humans are supposed to decide what kind of world humans and nonhumans should live in? However, there are so many people on Earth that adjudication seems no longer a matter of choice. Everything people do has a consequence, an outcome, a repercussion. Humans play God by taking action. But by refusing to act when we have the capacity, we play God by default, so we may as well take action.

U.S. Magistrate Willard W. McEwen Jr. acquitted Rob Puddicombe on the two federal charges. Puddicombe escaped a year in prison, and he responded to the verdict, "I only regret that the animals on Anacapa didn't get the same fair trial I did."

But they did, I think, as I come down the trail into the campground. According to data provided by the nonprofit group Island Conservation, back in 2000, when black rats still pillaged seabird nests on Anacapa Island, there were nine nesting Xantus's murrelets. In 2005, after rats had been fully eradicated from the island, there were twenty-four nesting Xantus's murrelets with 90 percent hatching success.

I hear the hiss of the Whisperlite stove, the whoosh of white gas. Bruce has started dinner and climbed the hill above our tent site to look at birds. The foghorn trumpets from the lighthouse bluff as I kneel to examine a tattered and regurgitated balloon at my feet. The severely frayed edges resemble a sea anemone's tentacles and indicate that the balloon likely exploded at high altitude and then dropped onto the ocean surface, where a gull ingested it. The orange Mylar gloms onto the dirt with a substance like tar. I peel it off and press it into my notebook.

When I reach Pinniped Point, a ranger, the first one I've seen since we arrived on Anacapa, greets me. He tells me that he and other park service officials released ten young bald eagles a few days ago on Santa Cruz Island. The eaglets hatched in Alaska, where bald eagles are so common people refer to them as "white-headed crows." "We'll see how they do," the ranger says. The reintroduction of bald eagles has helped shape the method in which pigs are hunted by ProHunt New Zealand. No poisons, snares, or

A juvenile bald
eagle tagged for
reintroduction
to the northern
Channel Islands

lead bullets may be used to hunt the pigs, in case the reintroduced bald
eagles should feed on the island's piled-up, rotting pig carcasses.

Later in the week, Bruce and I paddle away from Anacapa and return
to Santa Cruz Island. Somewhere near Portuguese Rock, off the west islet,

my paddle thumps the bloated body of a California harbor seal. Belly-up, its skin sunburned and scarred, the dead seal's blubber has been some scavenger's meal. I hope the seal carcass, possibly contaminated with DDT, doesn't attract any bald eagles, because the loss of one bald eagle remains something neither the species nor the chain of integrated Channel Island communities can afford.

In his book *Feral Future*, biologist and writer Tim Low focused on the problem of Australia's introduced species and addressed the need for action: "We will never evict all the pests we already have, but we can try much harder to keep new ones out. A new ecology is emerging, one we don't yet understand, but one that will debase the marvellously rich diversity of life on earth unless we manage it well."

What will this "new ecology" look like? The thought of rats, pigs, and deer mice shuddering in the throes of death is upsetting, but without a concerted effort to remove the black rats and other invasive species around the world, more extinctions of native animals are inevitable.

Epilogue   Chasing Bees

A dry leaf on a fast ride down the drain,
I rained fruitjarsful of pebbles,
built a mud dam,
stirred a spout to suck it in.
How far could one raft hold its own?
Oh, pretty far I want to say now
remembering how in spite of me
it righted and kept going.

—Kathryn St. Thomas, "Child's Play"

My friend Elaine Evans is an entomologist who studies bumblebees of the species *Bombus impatiens*. Along with six other entomologists, she painted a total of 1500 worker bees with white Testor's Model Master paint in order to spot them in the field and observe how far the female workers were willing to forage from the nest.

I visited her research site on the U.S. Army Reserve Center in St. Paul, Minnesota, following Elaine along the dirt roads inside the complex, which had been built on top of a toxic waste dump. While she chased bees, I admired the sticky yellow gumweed flowers that grew as corrupted exhalations, out-breaths of junk soil and slag.

Marching along in her wide-brim beekeeper's hat, backpack joggling loose around her shoulders, Elaine hooted whenever she spotted a painted worker bee whose hind legs hung heavy with orange pollen, as if carrying overfull saddlebags. "Oh, there she goes. No, wait. I lost her. No, there she is!" Like Elaine, I toted pad and pencil, but instead of ticking off the numbers of painted bees, I noted some of the plant invaders common to the area: garlic mustard, spotted knapweed, and white and yellow sweet clover.

Coming up with solutions to the problem of invasive species is a lot like chasing painted bumblebees. When you think you've found one, it may turn out not to be what you're looking for, so you have to persist, keeping the goal in mind while staying alert and dedicated. This book has aimed to do more than reduce the number of people who might confront an amnesty bin or find themselves seated on a plane 30,000 feet in the air, as wide-eyed and unsuspecting as that man who tried to take his pet snake to Hawaii. My hope is that readers will begin to take action to slow the spread of exotic species on behalf of the thousands of islands that do not require visitors to fill out declaration forms and do not provide amnesty bins.

Elaine was driven, darting here and there, tiptoeing up to a bush, and then dashing off after yet another worker bee. "I can't help it," she said. "I want to solve this puzzle. These bees work for the good of their colony, not themselves, and I want to know how far they'll go to contribute to the communal honey pot."

How far are people willing to go to conserve native species? "A long way," said Ken Owens of the Santa Cruz Island Restoration Project. "People want to know about this stuff," he told me. "They're eager to be part of the restoration process, but we've got to educate the general public. Right now, the problem of invasive species still gets tossed around inside scientific circles and not enough information gets out to island visitors." Ken has taken a pragmatic approach to this problem by talking with visitors during their hour-long boat ride to Santa Cruz Island, and by installing boot brushes on the mainland and on the island so visitors can scrape seeds and mud off their shoes and socks.

Owens has noticed that most of the plant invaders on Santa Cruz have sprouted up around the campgrounds and along the edges of hiking trails. He encourages people not to hike through weed-infested areas, but to stay on trails to avoid disturbing soils and making it easier for invasive plants to

A crowd of visitors receiving an orientation to Santa Cruz Island

establish. "People don't realize they need to look at their feet and examine their tents, sleeping bags, and pads," Owens remarked. "Caked mud on the soles of shoes can carry the pathogen that causes sudden oak death, a disease not limited to oak trees." As ethnobiologist Patty West pointed out, checking personal items for potential invaders is only one part of a multifaceted approach to halting the spread of invasive species, but it's an important one that involves everyone.

The dedicated work of scientists and resource managers is also crucial

Shoes covered
with the seeds
of exotic species

to the process of eradicating invaders. I remember talking to a park service maintenance worker on Santa Cruz Island as he stood chest-deep in yellow star thistle. "I'm getting rid of this darn weed. It's going, one way or another," he told me, in the upbeat voice of someone who believes anything is possible if you just try hard enough. With gloved hands, he pulled out one plant by its roots, and then another and another.

Kate Faulkner echoed that determination and affirmed the practical view of many other resource managers I spoke to: "Restoring native balance doesn't mean going back to some time before Europeans arrived, some pre-conquistador ecology. That's impossible," she said. "Restoration means tipping the scales back in favor of native plants and animals. It

Boot brushes on the mainland pier provide an opportunity for visitors to remove invasive seeds from their footwear before their trip to Santa Cruz Island.

means restoring native ecological processes to the best level we can and supporting the continuance of those processes."

Along with scientists and resource managers, regular people can address these problems. At home, we can learn to identify and remove invaders from our yards. Using lists from our local agricultural extension service, we can replace them with native plants. Rather than ordering potentially harmful plant material on the Internet, we can shop at local nurseries, and we can request that they carry native plants and refrain from selling exotics known to escape the confines of their landscaped environments.

In our communities, we can develop campaigns for public awareness of the invasive species problems in our own regions. We can start a native plant garden to demonstrate that "ornamental" doesn't have to mean non-native. We can communicate with urban- and rural-development planners to encourage them to plant native species, and we can call our state high-way departments to let them know that we want them to plant native species exclusively.

In wild areas, we can keep pets from running loose, to ease stress on wildlife and reduce soil disturbance. According to cheatgrass researcher Kim Allcock, those who ride horses or use pack animals in the backcountry should use certified weed-free feed. If it isn't immediately available, customers should keep asking their feed stores to stock it; with sufficient consumer demand, the feed industry will respond by producing a clean product.

We can refuse to support the exotic pet trade by forgoing ownership of exotic animals. Contributing to organizations like Three Ring Ranch Exotic Animal Sanctuary in Hawaii, a refuge for exotic animals brought to Hawaii and confiscated or abandoned, will help it to continue its work. This nonprofit facility, licensed by the U.S. Department of Agriculture to house exotics, is an amnesty site, meaning that no one who brings an animal to the sanctuary is prosecuted. The animals are rehabilitated and kept, rather than released, to keep them away from Hawaii's native wildlife. The sanctuary also offers education programs for children and adults to alert them to the problems caused by exotic-pet ownership.

When traveling, we can exercise caution by not carrying animals, fruits, or grains across borders unless we go through proper quarantine procedures. We can clean off recreational boats and blow out bilge water to avoid carrying aquatic species from one waterway to another. Not dumping live bait into the water after fishing or releasing aquarium fish or other aquatic animals into waterways are two other actions we can take.

Volunteer opportunities abound for people who can give their time and labor. In fact, all the islands I visited rely on volunteers to varying degrees. Ash Meadows sponsors Volunteer Days, the only time when people working to restore the refuge can legally swim in the pools. Pupfish thrive in fast-moving water, whereas the invasive crawfish prefer sluggish, even still pools. While enjoying the 84°F water, volunteers wield linoleum knives, cutting back the year's growth of cattails that choke the fast-moving flows pupfish thrive in. Their work helps to restore the refuge, something three staff members alone cannot do.

Other volunteer opportunities might be generated using models already in place. For example, in Hawaii the Oahu Invasive Species Committee provides ample opportunities for direct community involvement in eradication and restoration projects. Volunteers can participate in service trips to survey areas for particular invasive plant species, while learning to identify the species and determine their abundance.

Multiple-agency task forces such as the Invasive Species Advisory Board in Arizona and the Coordinating Group on Alien Pest Species in Hawaii bring together more than fifteen organizations and agencies, including the Department of Agriculture, U.S. Customs, and the Postal Service, to address invasive species impacts on health, the economy, and the environment. Eventually these groups may seek citizen participation in much the same way former park service superintendent Tim Setnicka envisioned.

In the larger political arena, we can use our power as voters to influence public policy. Lack of money for resource management is squelching the ability of many refuges and parks to eradicate invaders. We can lobby for more restoration money and support legislators who champion the protection of native species. We can also support legislation that regulates the traffic of plants and animals through more stringent transport guidelines and quarantine measures in our various states.

Bernie Tershy suggested that people give money to organizations that provide public education and do on-the-ground conservation work to eradicate invasive species. Island Conservation and Grupo de Ecología y Conservación de Islas are two such groups working in the Sea of Cortés (a list of others can be found through a link on the University of Arizona Press Website, http://www.uapress.arizona.edu/books/bid1762.htm). Tershy also encouraged writing letters to the editors of local newspapers and visiting some of the islands to appreciate the impact that eradication of invasive species has had on the native wildlife. "Take Anacapa Island," he said. "Seabird populations nearly doubled the year after the black rat was eradicated." The recovery of Anacapa's seabird population is one of many dramatic comebacks demonstrating our ability to restore native species and their habitats.

Elaine's bumblebees traveled a long way while collecting nectar and pollen from the native plants, such as wild bergamot, remaining on the reserve. She told me that farmers and the rest of us benefit from these bumblebees because they pollinate crops we rely on, like tomatoes, eggplants, peppers, melons, and many types of berries. The bees work hard for the good of their colonies—and incidentally for us—and their populations are only as healthy as the habitat they live in.

When I taught island ecology to sixth graders, we reviewed the definition of a habitat as a place that has all the necessities for maintaining life: shelter, food, water, space, and sunlight. Then I gave the children each a penny and told them that it represented their island, and they needed to build a habitat on their coin, a surface just three-quarters of an inch in diameter. The kids ran around placing blades of grass or seeds on their pennies to symbolize food; twigs, shells, or pebbles to represent shelter; and so on. A penny, like an island, has only so much space. The children soon understood what an introduced species could do to their penny habitats, the copper islands in their palms.

Appendix List of Species

Amargosa pupfish (*Cyprinodon nevadensis mionectes*)
American oystercatcher (*Haematopus palliatus*)
American white pelican (*Pelecanus erythrorhynchos*)
Anacapa deer mouse (*Peromyscus maniculatus anacapae*)
Anise swallowtail butterfly (*Papilio zelicaon*)
Asian long-horned beetle (*Anoplophora glabripennis*)
Athel (*Tamarix aphylla*)
Australian saltbush (*Atriplex semibaccata*)

Bald eagle (*Haliaeetus leucocephalus*)
Big sagebrush (*Artemisia tridentata*)

Black rat (*Rattus rattus*)
Bladder parsnip (*Lomatium utriculatum*)
Blue-footed booby (*Sula nebouxii*)
Blue whale (*Balaenoptera musculus*)
Boojum tree (*Fouquieria columnaris*)
Brown booby (*Sula leucogaster*)
Brown tree snake (*Boiga irregularis*)
Buckwheat (*Eriogonum* spp.)
Buffelgrass (*Pennisetum ciliare*)
Bumblebee (*Bombus impatiens*)

California brown pelican (*Pelecanus occidentalis californicus*)
California gull (*Larus californicus*)
California harbor seal (*Phoca vitulina*)
California sea lion (*Zalophus californianus*)
Camel (*Camelops* spp.)
Cardón tree (*Pachycereus pringlei*)
Cenizo (*Leucophyllum frutescens*)
Chain-link cholla (*Cylindropuntia imbricata*)
Cheatgrass (*Bromus tectorum*)
Chuckwalla (*Sauromalus* spp.)
Claret-cup cactus (*Echinocereus triglochidiatus*)
Clark's grebe (*Aechmophorus clarkii*)
Common dolphin (*Delphinus delphius*)
Compass barrel cactus (*Ferocactus cylindraceus*)
Convict cichlid (*Cichlasoma nigrofasciatum*)
Cortés halibut (*Paralichthys aestuarius*)
Cortés round stingray (*Urolophus maculatus*)
Costa's hummingbird (*Calypte costae*)
Cottonwood (*Populus deltoides wislizenii*)
Cow parsnip (*Heracleum lanatum*)

Creosote bush (*Larrea divaricata*)
Crested wheatgrass (*Agropyron cristatum*)
Crystal iceplant (*Mesembryanthemum crystallinum*)
Cui-ui (*Chasmistes cujus*)
Cutthroat trout (*Oncorhynchus clarkii clarkii*)

Deer mouse (*Peromyscus maniculatus*)
Desert rosy boa (*Lichanura trivirgata gracia*)
Domestic cat (*Felis catus*)
Double-crested cormorant (*Phalacrocorax auritus*)

Eucalyptus tree (*Eucalyptus globulus*)
European wild boar (*Sus scrofa*)

Feral pig (*Sus scrofa*)
Florida scrub jay (*Aphelocoma coerulescens*)
Fourwing saltbush (*Atriplex canescens*)
Foxtail barley (*Hordeum jubatum*)

Garibaldi fish (*Hypsypops rubicundus*)
Garlic mustard (*Alliaria petiolata*)
Giant coreopsis or tree sunflower (*Coreopsis gigantea*)
Giant manta ray (*Manta birostris*)
Giant reed (*Arundo donax*)
Golden eagle (*Aquila chrysaetos*)
Greasewood (*Sarcobatus vermiculatus*)
Gumweed (*Grindelia fraxinipratensis, Grindelia squarrosa, Grindelia stricta*)

Heermann's gull (*Larus heermanni*)
Hoffmann's rock cress (*Arabis hoffmannii*)
Horse (*Equus pacificus*)

Hottentot fig (*Carpobrotus edulis*)
House mouse (*Mus musculus*)

Island barberry (*Berberis pinnata*)
Island gray fox (*Urocyon littoralis*)
Island scrub jay (*Aphelocoma insularis*)
Island scrub oak (*Quercus pacifica*)

Jojoba bush (*Simmondsia chinensis*)
Jumbo squid (*Dosidicus gigas*)

Kangaroo rat (*Dipodomys* spp.)

Largemouth bass (*Micropterus salmoides*)
Loggerhead sea turtle (*Caretta caretta*)
Louisiana or red swamp crawfish (*Procambarus clarkii*)

Merriam's kangaroo rat (*Dipodomys merriami*)
Monarch butterfly (*Danaus plexippus*)
Moon jellyfish (*Aurelia aurita*)

Nasturtium (*Tropaeolum majus*)
Norway rat (*Rattus norvegicus*)
Nuttall's larkspur (*Delphinium nuttallianum*)

Ocotillo (*Fouquieria splendens*)
Old man cactus (*Lophocereus schottii*)
Osprey (*Pandion haliaetus*)
Owens pupfish (*Cyprinodon radiosus*)

Pacific slender salamander (*Batrachoseps pacificus*)
Pickleweed (*Salicornia* spp.)
Piñon pine (*Pinus edulis*)
Pister's pupfish (*Cyprinodon pisteri*)
Pister's snail (*Pyrgulopsis pisteri*)
Pitaya agria (*Stenocereus gummosus*)
Prince's plume (*Stanleya pinnata*)

Rabbitbrush (*Ericameria nauseosa*)
Rainbow trout (*Oncorhynchus mykiss*)
Red brome (*Bromus madritensis rubens*)
Reddish egret (*Egretta rufescens*)
Red-eared slider (*Trachemys scripta elegans*)
Red mangrove (*Rhizophora mangle*)
Red pelagic crab (*Pleuroncodes planipes*)
Red sea urchin (*Strongylocentrotus franciscanus*)
Red-shouldered hawk (*Buteo lineatus*)
Rock dove (*Columba livia*)
Russian thistle (*Salsola tragus*)

Sahara mustard (*Brassica tournefortii*)
Salt Creek pupfish (*Cyprinodon salinus salinus*)
Salt grass (*Distichlis stricta*)
Salt grass skipper (*Polites sabuleti*)
Santa Cruz Island buckwheat (*Eriogonum arborescens*)
Santa Cruz Island live-forever (*Dudleya nesiotica*)
Screwbean mesquite (*Prosopis pubescens*)
Shadscale (*Grayia spinosa*)
Siberian wheatgrass (*Agropyron fragile*)
Side-blotched lizards (*Uta* spp.)
Southwestern willow flycatcher (*Empidonax trailii extimus*)

Spiny chuckwalla (*Sauromalus hispidus*)
Spiny saltbush (*Atriplex confertifolia*)
Spotted knapweed (*Centaurea maculosa*)
Sweet fennel (*Foeniculum vulgare*)

Tamarisk (*Tamarix chinensis*)
Totoaba (*Totoaba macdonaldi*)

Vaquita (*Phocoena sinus*)
Velvet ash (*Fraxinus velutina* var. *coriacea*)
Vietnamese potbellied pig (domestic breed of *Sus scrofa*)

Water boatmen (*Corixa* spp.)
Western diamondback rattlesnake (*Crotalus atrox*)
Western grebe (*Aechmophorus occidentalis*)
Western gull (*Larus occidentalis*)
Western mosquito fish (*Gambusia affinis*)
Western rattlesnake (*Crotalus viridis*)
Western scrub jay (*Aphelocoma californica*)
Wheat (*Triticum* spp.)
White sweet clover (*Melilotus alba*)
Wild bergamot (*Monarda fistulosa*)
Willow (*Salix* spp.)
Winterfat (*Krascheninnikovia lanata*)

Xantus's murrelet (*Synthliboramphus hypoleucus*)

Yellow star thistle (*Centaurea solstitialis*)
Yellow sweet clover (*Melilotus officinalis*)

Zebra mussel (*Dreissena polymorpha*)

Acknowledgments

I received a great deal of help from resource managers and scientists, who were generous with their time and knowledge of island habitats. I am grateful to them all. Some readers have met in the book. Many others, not named here, also contributed time and information. Donna Withers and Styron Bell taught me something new during every trip to Anaho Island. I am indebted to Phil Pister, who graciously shared his knowledge from many years of working with pupfish. His good humor, friendship, and careful reading of parts of this manuscript helped immensely. I thank Patty West, who offered her expert knowledge of the Bahía de los Angeles archipelago. Kate Faulkner gave freely of her time and her experience restoring the Channel Islands. Rick Moser worked tirelessly to produce the maps for this book. Crystal Atamian, Kerry Grimm, and Bruce McKenzie were exceptional companions to explore the islands with.

Others who kindly furnished important information include Kim Allcock, Dan Anderson, Ted Angle, Kyle Ashton, Charles Barclay, Steve Beaupre, staff of the California Department of Transportation, Harry Carter, Ruark Cleary, Bruce Eilerts, Guillermo Galván, Linda Greene, John Hall, Herman Hill, Ann Huston, Richard Mack, Michael Markarian, Dave Marshall, Robert Nowak, Ken Owens, Antonio Resendiz, Don Sada, Randy Smith, Robin Tausch, Bernie Tershy, Nancy Vivrette, Dieter Wilken, and Vicki Wolfe.

This book is for my mother, the poet Kathryn St. Thomas, who offered constant encouragement and editing advice. A million thanks would not be enough to show my appreciation. I've had the privilege of working with editor Patti Hartmann, who believed in this book and kept me on track. Finally, I thank all of the reviewers who read and commented on the manuscript; their suggestions were invaluable.

Bibliography

Anderson, John. 1990. *Enememe's Friends: Chumash Theology*. Santa Fe, N.M.: Center for Indigenous Arts and Cultures Press.

Anderson, John. 1996. *A Circle within the Abyss: Chumash Metaphysics*. 3rd ed. Santa Fe, N.M.: Center for Indigenous Arts and Cultures Press.

Aridjis, Homero. 2001. War on nature in La Paz. *Reforma* 4 March.

Aridjis, Homero. 2001. Baja in decline (Baja a la baja). *Reforma* 14 October.

Bakker, Elna. 1984. *An Island Called California: An Ecological Introduction to Its Natural Communities*. 2nd ed. Berkeley: University of California Press.

Barthell, J. F., J. M. Randall, R. W. Thorp, and A. M. Wenner. 2001. Promotion of seed set in yellow star-thistle by honey bees: evidence of an invasive mutualism. *Ecological Applications* 11(6): 1870–1883.

Berger, Bruce. 1998. *Almost an Island: Travels in Baja California*. Tucson: University of Arizona Press.

Booth, Jerry. 2002. NRS research that revealed collapse of island fox population now focuses on halting species extinction. *Transect* 20(2): 2–5.

Bright, Christopher. 1998. *Life out of Bounds: Bioinvasion in a Borderless World*. New York: W. W. Norton & Co.

Bright, Christopher. 1999. Invasive species: pathogens of globalization. *Foreign Policy* 116: 50–64.

Broderick, Eileen. 2003. Aquarium escapees. Online posting, Alien Species Discussion Group. Accessed 2 and 4 May. http://aliens-l@indaba.iucn.org.

Browne, J. Ross. 1868. Explorations in Lower California. *Harper's New Monthly Magazine* 38(223): 9–24.

Browning, Robert. 1842. *The Pied Piper of Hamelin: A Child's Story*. 1910. Chicago: Rand, McNally Co.

Carson, Rachel. 1962. *Silent Spring*. 1994. New York: Houghton Mifflin.

Christensen, Jon. 2000. Fire and cheatgrass conspire to create a weedy wasteland. *High Country News* 32(10).

Chumash ruckus at fiesta. 2002. *The Santa Barbara Independent* 8 August: 16.

Courtenay, W. R., Jr., and G. K. Meffe. 1989. Small fishes in strange places: a review of introduced poeciliids. Pp. 319–331 in *Ecology and Evolution of Livebearing Fishes (Poeciliidae)*. G. K. Meffe and F. F. Snelson Jr., eds. Englewood Cliffs, N.J.: Prentice Hall.

Cox, George W. 1999. *Alien Species in North America and Hawaii: Impacts on Natural Ecosystems*. Washington, D.C.: Island Press.

Cravalho, Domingo, and Fred Kraus. 2001. The risk to Hawai'i from snakes. *Pacific Science* October: 409–417.

Cronk, Quentin C. B., and Janice L. Fuller. 1995. *Plant Invaders: A 'People and Plants' Conservation Manual*. London: Chapman and Hall.

Crosby, Alfred W. 1972. *The Columbian Exchange: Biological and Cultural Consequences of 1492*. Westport, Conn.: Greenwood Press.

Crosby, Alfred W. 1986. *Ecological Imperialism: The Biological Expansion of Europe, 900–1900*. Cambridge: Cambridge University Press.

Crosby, Harry W. 1984. *The Cave Paintings of Baja California*. La Jolla, Calif.: Copley Books.

D'Antonio, Carla M. 2000. Fire, plant invasions, and global changes. Pp. 65–93 in *Invasive Species in a Changing World*. Harold A. Mooney and Richard J. Hobbs, eds. Washington, D.C.: Island Press.

Darwin, Charles. 1860. *The Voyage of the Beagle*. 1962. Leonard Engel, ed. New York: Doubleday.

De Poorter, Maj. 2003. Glossary/definitions. Online posting, Alien Species Discussion Group. Accessed 12 February. http://aliens-l@indaba.iucn.org.

De Ruff, Robert. 2003. Plants of Upper Newport Bay, California. Accessed 19 April. http://mamba.bio.uci.edu/~pjbryant/biodiv/PLANTS2/Aizoaceae/Mes embryanthemum_crystallinum.htm.

Dukes, Jeffrey S., and Harold A. Mooney. 1999. Does global change increase the success of biological invaders? *Trends in Ecology and Evolution* 14(4): 135–139.

Ehrlich, Gretel. 1991. *Islands, the Universe, Home*. New York: Penguin.

Elton, Charles S. 1958. *The Ecology of Invasions by Animals and Plants*. 2000. Chicago: University of Chicago Press.

Evans, Roger M., and Fritz L. Knopf. 1993. American white pelican. *The Birds of North America* 57: 1–24.

Flaherty, Mary. 2004. East Bay: migration of monarch butterflies gets naturalists aflutter. *San Francisco Chronicle* 17 December: B2, B5.

Foster, Steven, and Christopher Hobbs. 2002. *A Field Guide to Western Medicinal Plants and Herbs*. Peterson Field Guide Series. Boston: Houghton Mifflin.

Fox, Douglas. 2003. Using exotics as temporary habitat: an accidental experiment on Rodrigues Island. *Conservation in Practice* 4(1): 32–37.

Francé, Raoul Heinrich. 1905. *Germs of Mind in Plants*. Chicago: Charles H. Kerr.

French, Hilary. 2000. *Protecting the Planet in the Age of Globalization*. Washington, D.C.: Worldwatch Institute.

Freudenberg, William R. 2004. Eagles, foxes, pigs, and people. Lecture. University of California, Santa Barbara. 12 October.

Garrison, Tom. 2002. *Essentials of Oceanography*. Pacific Grove, Calif.: Brooks/Cole.

Grant, Campbell. 1974. *Rock Art of Baja California*. Los Angeles: Dawson's Book Shop.

Greenlaw, Linda. 2002. *The Lobster Chronicles: Life on a Very Small Island*. New York: Hyperion.

Homer. 1997. *Iliad*. Stanley Lombardo, trans. Indianapolis: Hackett.

Hubbel, Sue. 1993. *Broadsides from the Other Orders: A Book of Bugs*. New York: Houghton Mifflin.

Jackson, Donald, and Mary Lee Spence, eds. 1970. *The Expeditions of John Charles Frémont: Travels from 1838–1844*. Vol. 1. Urbana: University of Illinois Press.

Junak, Steve. 1995. *A Flora of Santa Cruz Island*. Santa Barbara, Calif.: Santa Barbara Botanic Garden.

Kareiva, Peter. 1996. Developing a predictive ecology for non-indigenous species and ecological invasions. *Ecology* 77(6): 1651–1652.

Keane, Ryan. 2003. Aquarium escapees. Online posting, Alien Species Discussion Group. Accessed 2 and 4 May. http://aliens-l@indaba.iucn.org.

Keesee, John, and Richard W. Rust. 1999. *A Student's Field Guide to Ash Meadows N.W.R.* Published by the authors.

Kelly, David. 2002. Animal activist finds himself in rat's nest of legal trouble. *Los Angeles Times* 15 December: B1.

Kettmann, Matt. 2003. Death for life on Anacapa Island. *The Santa Barbara Independent* 29 April: 12.

Knack, Martha C., and Omer C. Stewart. 1999. *As Long as the River Shall Run*. 2nd ed. Berkeley: University of California Press.

Krutch, Joseph Wood. 1961. *The Forgotten Peninsula: A Naturalist in Baja California*. 1986. Tucson: University of Arizona Press.

Lemly, Dennis A. 1994. Agriculture and wildlife: ecological implications of subsurface irrigation drainage. *Journal of Arid Environments* 28: 85–94.

Leopold, Aldo. 1949. *A Sand County Almanac and Sketches Here and There*. London: Oxford University Press.

Leopold, Aldo. 1953. *Round River: From the Journals of Aldo Leopold.* 1993. Luna B. Leopold, ed. New York: Oxford University Press.

Low, Tim. 1999. *Feral Future: The Untold Story of Australia's Exotic Invaders.* Ringwood, Victoria: Penguin Books.

Low, Tim. 2004. *The New Nature: Winners and Losers in Wild Australia.* Sidney: Penguin Books.

Mack, Richard N. 2002. Eradicating invasive plants: hard-won lessons for islands. Pp. 164–172 in *Turning the Tide: The Eradication of Invasive Species.* D. Veitch and M. Clout, eds. Auckland, New Zealand: Invasive Species Specialty Group of the World Conservation Union.

Mack, Richard N. 2003. Phylogenetic constraint, absent life forms and pre-adapted alien plants: a prescription for biological invasions. *International Journal of Plant Sciences* 164(Suppl. 3): S185–S196.

Mack, Richard N. 2003. Plant naturalizations and invasions in the eastern United States, 1634–1860. *Annals of the Missouri Botanical Garden* 90: 77–90.

Mack, Richard N. 2003. Global plant dispersal, naturalization, and invasion: pathways, modes, and circumstances. Pp. 3–30 in *Global Pathways of Biotic Invasions.* G. Ruiz and J. Carlton, eds. Washington, D.C.: Island Press.

Mann, Charles C. 2002. 1491. *The Atlantic Monthly* 289(3): 41–52.

Marshall, David B., and Leroy W. Giles. 1953. Recent observations on birds of Anaho Island, Pyramid Lake, Nevada. *Condor* 55: 105–116.

McCrary, M. D., and P. H. Bloom. 1984. Lethal effects of introduced grasses on red-shouldered hawks. *Journal of Wildlife Management* 48(3): 1005–1008.

McNeely, J. A., Laurie E. Neville, and Marcel Rejmánek. 2003. When is eradication a sound investment? Strategically responding to invasive alien species. *Conservation in Practice* 4(1): 30–31.

Meighan, Clement W., and V. L. Pontoni, eds. 1978. *Seven Rock Art Sites in Baja California.* Socorro, N.M.: Ballena Press.

Mellink, Eric. 2002. Invasive vertebrates on islands of the Sea of Cortés. Pp. 112–125 in *Invasive Exotic Species in the Sonoran Region.* Barbara Tellman, ed. Tucson: University of Arizona Press and Arizona-Sonora Desert Museum.

Mercade, Jose A., ed. 1988. *Bahía de los Angeles Educational Resources Series*. No. 1. Glendale, Calif.: Glendale Community College.

Minckley, W. L., ed. 1973. *Fishes of Arizona*. Phoenix: Arizona Game and Fish Department.

Minckley, W. L., and James Deacon, eds. 1991. *Battle against Extinction: Native Fish Management in the American West*. Tucson: University of Arizona Press.

Mooney, Harold A. A., and Richard J. Hobbs, eds. 2000. *Invasive Species in a Changing World*. Washington, D.C.: Island Press.

Moore, Michael. 1989. *Medicinal Plants of the Desert and Canyon West*. Santa Fe: Museum of New Mexico Press.

Nabhan, Gary. 2000. Cultural dispersal of plants and reptiles to the Midriff Islands of the Sea of Cortés: integrating indigenous human dispersal agents into island biogeography. *Journal of the Southwest* 42(3): 545–558.

Nabhan, Gary. 2003. *Singing the Turtles to Sea: The Comcáac (Seri) Art and Science of Reptiles*. Berkeley: University of California Press.

Nápoli, Ignacio Maria. 1970. *The Cora Indians of Baja California*. James Robert Morriarty III and Benjamin F. Smith, eds. Los Angeles: Dawson's Book Shop.

National Research Council. 2002. *Predicting Invasions by Nonindigenous Plants and Plant Pests*. Washington, D.C.: National Academy of Sciences.

Newfield, Melanie. 2003. Reckless, irresponsible or monumentally stupid? Online posting, Alien Species Discussion Group. Accessed 4 May. http://aliens-l@indaba.iucn.org.

Norris Brenzel, Kathleen, ed. 2001. *Sunset Western Garden Book*. 7th ed. Menlo Park, Calif.: Sunset.

O'Dell, Scott. 1960. *Island of the Blue Dolphins*. 1990. Boston: Houghton Mifflin.

Oliver, Mary. 1992. *New and Selected Poems*. Boston: Beacon Press.

Oliver, Rice D. 1993. *Lone Woman of Ghalas-hat*. Tustin, Calif.: California Weekly Explorer.

Orwell, George. 1950. Politics and the English language. *Shooting an Elephant and Other Essays*. 1984. New York: Harcourt, Brace & Co.

Parker, Clifton B. 2003. The aliens among us. *U.C. Davis Magazine* 20(3): 22–27.

Parker, I. M., D. Simberloff, W. M. Lonsdale, K. Goodell, M. Wonham, P. M. Kareiva, M. W. Williamson, V. Holle, P. B. Moyle, J. E. Byers, and L. Goldwasser. 1999. Impact: toward a framework for understanding the ecological effects of invaders. *Biological Invasions* 1: 3–19.

Parker, William, and William S. Brown. 1974. Mortality and weight changes of Great Basin rattlesnakes (*Crotalus viridis*) at a hibernaculum in northern Utah. *Herpetologica* 30(3): 234–239.

Perrault, Anne, Carroll Muffet, Stas Burgiel, Morgan Bennett, and Aimee Delach. 2003. Invasive species, agriculture, and trade: case studies from the NAFTA context. Paper presented at the Second North American Symposium on Assessing the Environmental Effects of Trade, Mexico City, 25–26 March.

Pister, Edwin P. 1993. Species in a bucket. *Natural History* 102(1): 14–17.

Pister, Edwin P. 1995. Fishes of the California Desert Conservation Area. Pp. 285–303 in *The California Desert: An Introduction to Natural Resources and Man's Impact.* Vol. 2. June Latting and Peter G. Rowlands, eds. Riverside, Calif.: June Latting Books.

Primack, Richard B. 1998. *Essentials of Conservation Biology.* 2nd ed. Sunderland, Mass.: Sinauer Associates.

Quammen, David. 1998. Planet of weeds. *Harper's Magazine* October: 57–69.

Radford, Tim. 2005. Two-thirds of the world's resources "used up." *Guardian Unlimited* 30 March 2005. Available via http://www.guardian.co.uk/international/story/0,3604,1447869,00.html.

Ragland, Jennifer. 2003. Rare bird hatches a comeback. *Los Angeles Times* (Ventura County ed.) 2 June: B1.

Rejmánek, Marcel, and David Richardson. 1996. What attributes make some plant species more invasive? *Ecology* 77(6): 1665–1661.

Richardson, David M., Petr Pyšek, Marcel Rejmánek, Michael G. Barbour, F. Dane Panetta, and Carol J. West. 2002. Naturalization and invasion of alien plants: concepts and definitions. *Diversity and Distributions* 6(2): 93–107.

Roberts, Norman C. 1989. *Baja California Plant Field Guide.* La Jolla, Calif.: Natural History Publishing.

Romano-Lax, Andromeda. 1993. *Adventure Kayaking: Baja*. Berkeley, Calif.: Wilderness Press.

Rusco, Elmer R., ed. 2000. *A Reporter at Large: Dateline—Pyramid Lake, Nevada*. Reno: University of Nevada Press.

Sacks, Oliver. 1996. *The Island of the Colorblind*. New York: Alfred A. Knopf.

Schaller, G. B. 1964. Breeding behavior of the white pelican at Yellowstone Lake, Wyoming. *Condor* 66: 3–23.

Scheibe, John. 2002. Suit over rat poison on island is dropped. *The Ventura County Star* (Ventura County ed.) 21 August: A3.

Schlag-Mendenhall, Matt. 2001. Plight of the pelican. *Birder's World* April: 42–47.

Schoenherr, Allan A., Robert C. Feldmeth, and Michael J. Emerson. 1999. *Natural History of the Islands of California*. Berkeley: University of California Press.

Scoppettone, Gary G., and Peter H. Riser. 2000. Reproductive longevity and fecundity associated with nonannual spawning in cui-ui. *Transactions of the American Fisheries Society* 129: 658–669.

Seacology. 2004. Preserving island environments and cultures. Accessed 5 January. http://www.seacology.com/facts/index.html.

Setnicka, Tim J. 2005. Ex–park chief calls for moratorium on island "hunt." *Santa Barbara News-Press* 25 March: A13.

Simberloff, Daniel. 2000. Foreword. Pp. vii–xiv in *The Ecology of Invasions by Animals and Plants*. By Charles Elton. Chicago: University of Chicago Press.

Soulé, Michael, and Gary Lease, eds. 1995. *Reinventing Nature? Responses to Postmodern Deconstruction*. Washington, D.C.: Island Press.

Spellenberg, Richard. 1979. *National Audubon Society Field Guide to North American Wildflowers (Western Region)*. New York: Chanticleer Press.

Steinbeck, John. 1941. *The Log from the Sea of Cortez*. 1995. New York: Penguin Books.

Steinbeck, John. 1947. *The Pearl*. 1966. New York: Viking Press.

Stewart, Amy. 2005. *The Earth Moved: On the Remarkable Achievements of Earthworms*. Chapel Hill, N.C.: Algonquin Books.

Sullivan, T. J. 2000. Rat-poisoning program to be tested on Anacapa. *The Ventura County Star* (Ventura County ed.) 22 November: C2.

Talev, Margaret. 2001. Court lets U.S. kill island's rats. *Los Angeles Times* 30 November: B6.

Tellman, Barbara. 2002. Human introduction of exotic species in the Sonoran region. Pp. 25–46 in *Invasive Exotic Species in the Sonoran Region*. Barbara Tellman, ed. Tucson: University of Arizona Press and Arizona-Sonora Desert Museum.

Todd, Kim. 2001. *Tinkering with Eden: A Natural History of Exotics in America*. New York: W. W. Norton & Co.

Tuchman, Barbara W. 1978. *A Distant Mirror: The Calamitous 14th Century*. New York: Ballantine.

University of Southern Florida, College of Arts and Sciences. 2003. Crayfish. Accessed 22 May 2003. http://chuma.cas.usf.edu/~ming/AboutAstacin/crayfish.htm.

U.S. Census Bureau. Census 2000. Available via http://www.census.gov.

U.S. Fish and Wildlife Service. 2001. NV-contaminant exposure of white pelicans nesting at Anaho Island National Wildlife Refuge. *Environmental Contaminants Program On-Refuge Investigations Sub-Activity*. Final report, Project IN30.

U.S. Geological Survey. 2003. Nonindigenous aquatic species. Accessed 6 April. http://nas.er.usgs.gov/fishes/accounts/poecilii/ga_affin.html.

U.S. National Park Service. 2000. *Anacapa Island Restoration Project*. Accessed 12 October. http://www.nps.gov/chis/naturalresources/airp.html.

Urban Education Partnership. 2005. Urban garden butterflies of southern California. Accessed 10 February 2005. http://www.laep.org/uclasp/ISSUES/butterflies/urban_garden.html.

Van Devender, Thomas R. 2002. Deep history of immigration in the Sonoran Desert region. Pp. 5–24 in *Invasive Exotic Species in the Sonoran Region*. Barbara Tellman, ed. Tucson: University of Arizona Press and Arizona-Sonora Desert Museum.

Vogel, Kenneth P. 2002. The challenge: high-quality seed of native plants to ensure establishment. *Seed Technology* 24: 9–15.

Vogel, K. P., and R. A. Masters. 2001. Frequency grid: a simple tool for measuring grassland establishment. *Journal of Range Management* 54: 653–655.

Warshall, Peter. 2001. Green Nazis? Restoration ecology and invasive species. *Whole Earth* 106: 40–43.

Wells, Herbert George. 1898. *The War of the Worlds.* 1977. New York: Oxford University Press.

West, Patricia. 2002. Floral richness, phytogeography, and conservation on islands in Bahía de los Angeles, Mexico. MSc. thesis, University of Arizona.

West, Patricia, and Gary Paul Nabhan. 2002. Invasive plants: their occurrence and possible impact on the central Gulf Coast of Sonora and the Midriff Islands in the Sea of Cortés. Pp. 91–111 in *Invasive Exotic Species in the Sonoran Region.* Barbara Tellman, ed. Tucson: University of Arizona Press and Arizona-Sonora Desert Museum.

The Wilderness Act. Public Law 88-157, 88th Congress, S. 4, 3 September 1964.

Williams, Ted. 2002. America's largest weed. *Audubon Magazine* 1: 22–27.

Williams, Ted. 2002. The eucalyptus: sacred or profane? *High Country News* 2: 3–6.

Wilson, Edward O. 1996. *In Search of Nature.* Washington, D.C.: Island Press/Shearwater Books.

Woodbury, W. Verne. 1966. The history and present status of the biota of Anaho Island, Pyramid Lake, Nevada. MSc. thesis, University of Nevada, Reno.

Zavaleta, Erika. 2000. Valuing ecosystem services lost to *Tamarix* invasion in the United States. Pp. 261–300 in *Invasive Species in a Changing World.* Harold A. Mooney and Richard J. Hobbs, eds. Washington, D.C.: Island Press.

Zwinger, Ann Haymond. 1989. *The Mysterious Lands: A Naturalist Explores the Four Great Deserts of the Southwest.* New York: E. P. Dutton.

Index

Black Death, 162–63

black rat: description of, 161f;
eradication of, 159, 160, 162,
170, 171–77, 178, 189–90;
rehabilitation of, 10; spread of,
162, 163, 164–66; threat posed
by, 161–62

bladder parsnip, 12

blue gum eucalyptus, 144

blue whale, 136

boobies, 116

boojum, 92, 97

boot brushes, 183, 186f

borate, 52

Bota, Isla (Boot Island), 114

brodifacoum, 175

Brown, Edmund "Pat," 68f

brown booby, 105

Browne, J. Ross, 93

brown pelican, 104, 105, 124, 135

brown tree snake, 12

buckwheat, 165f, 168

buffelgrass, 77, 78, 121, 122

bumblebees, 182, 190

bunchgrass, 15, 38

butterflies, 12, 53–54, 145

Cabeza de Caballo, Isla (Horsehead
Island), 105

Cabrillo, Juan Rodríguez, 130, 131,
134–35

Calaveras, Isla (Skull Island), 116

California, El Gulfo de. *See* Cortés, Sea
of (Mexico)

California brown pelican, 19–20

California Dept. of Fish and Game, 61

California Dept. of Transportation
(CALTRANS), 78–80

California harbor seal, 128, 158, 170

California newt, 9

California sea lion, 139

California tree frog, 9

canine distemper, 11

carbon, applying to soil, 31

cardón trees, 93

Carson, Rachel, 137

Carter, Harry, 161–62

cats, 77–78, 100, 120

cattle, nonnative seeds spread by, 29

Cerraja, Isla (Lock Island), 113

chain-link cholla, 93

Channel Islands (Calif.): human
inhabitants of, 133, 146; location
and access to, 7, 7f, 129f;
management of, 132–33; native
species of, 127–28, 146–47, 180;
nonnative species of, 126, 135,
149–51; study of, 6

cheatgrass: adaptability of, 37; bird
population, impact on, 38; fires,
increase due to, 38–39, 40–41;
on hills and mountains, 15, 44;

63; native *versus* exotic species, impact on, 6; plant and tree, 184

dogs, 11, 100–101

dolphins, 136

drought tolerance, 77, 109

Earth-island analogy, 13–14

ecological pressures, resistance to, 6

economic impact of invasive species, 8, 86–87, 189

ecosystems, threats to, 5, 90–91, 157

electrofishing, 61

Endangered Species Act, 64, 69

environment, invasive species impact on, 189

environmental disasters, 41–42

Environmental Protection Agency, 137

eucalyptus, 142, 143–45, 144f, 146, 153

evolution, 50–51

excrement, disposal of, 107–8

exotic species: animals, 10–11, 16, 187; defined, 5; plants, 16

extinction, 6, 12, 14, 51

Fallon, Nevada, 25–26

Faulkner, Kate: on eagles, 139, 150–51; on eradication, 155, 162, 166, 172, 173, 175; on invasive plants, 145, 170; on restoration, 160, 185

feral animals, 100, 101. *See also* pigs, feral

fires, nonnative species role in spread of, 8, 38–39, 40–41, 42, 144–45

fish: death of, 22–23; desert, 65, 70; native, 9, 54, 67; nonnative, 9, 54, 57–59, 67, 69; refuges for, 54, 55

Fish Slough (Calif.): conservation at, 62f, 63; location and access to, 7, 7f, 47f; study of, 6, 48

Florida scrub-jay, 142

food chain, contaminants in, 138–39

fourwing saltbush, 43, 51, 87, 109

foxtail barley, 170

Frémont, John C., 26

Fund for Animals, The, 162, 174, 175

Gable, Clark, 18

Galván, Guillermo, 99, 101–2

garden nasturtium, 88

Garibaldi fish, 160

garlic mustard, 182

giant reed, 77

golden eagle, 147, 148f, 149–51, 158

grain shipments, contaminants in, 29, 39, 94

grazing animals, environmental impact of, 141–42

grazing animals, nonnative seeds spread by, 29

Great Basin, Nevada, fires in, 38–39

malaria control agents, 67

mammal extinctions, 6

Markarian, Michael, 175–76

McDowell fire, Scottsdale, Ariz., 1995,
41–42

McKelvey, Sharon, 58–59, 67, 69

Merriam's kangaroo rat, 111

Midriff Islands (Mexico):
conservation efforts on, 106–7;
development projects on, 84;
location and access to, 7, 7f, 74,
75f, 76f, 110; nonnative species
on, 76–78, 109, 113, 120, 121,
122–23; study of, 6; wildlife of,
111, 112, 123

military transport, invasive species
introduction through, 11–12

monarch butterfly, 145

Monkey Springs pupfish, 60

Monroe, Marilyn, 18

Montrose Chemical Corporation, 138,
139

moon jellies, 120

Morelos Diversion Dam, 84–85

mosquitofish, 9, 67

multiple-agency task forces, 188–89

Napolitano, Janet, 122

National Park Service: Channel Island
management by, 128, 140, 145,
169; pig eradication by, 127, 152,
155, 156; rat eradication by, 171,
172, 175, 176–77

Native American groups, 122

native plants and vegetation: consumer
selection of, 186–87; extinction
of, 6; growth, inhibition of, 13;
regeneration, obstacles to, 8;
replacing with exotics, 30; studies
on, 16; threats to, 144

native species: conservation of,
183; environmental adaptation
by, 12, 145; public education
regarding, 183–84, 187, 190;
reintroduction of, 31, 139, 158,
178–79; restoration of, 32, 42–43,
190; threats to, 5, 12–13, 160–61;
vulnerability of island, 5–6, 114,
190

native *versus* nonnative species: climate
change impact on, 6; competition
between, 8, 30, 31–32, 81, 86;
disease impact on, 6; drought
tolerance of, 109; rehabilitation
of, 9–10

natural herbivores, 6

natural predators, 6

Nature Conservancy, 121, 127, 140,
152, 155

Navarro, James, 136–37

Nevada, native species of, 49–50, 51

Nevada Dept. of Fish and Game, 61

About the Author

Ceiridwen Terrill worked as a volunteer interpreter for Cumberland Island National Seashore and Canyonlands National Park, as well as a wilderness ranger for the U.S. Forest Service in the Marble Mountain Wilderness of California. She received a PhD in English from the University of Nevada, Reno, with an emphasis in Literature and Environment. She is an assistant professor at Concordia University in Portland, Oregon, where she teaches literary nonfiction, including environmental journalism and science writing. Her recent essays have appeared in *Oxford American* and *Isotope: A*

Journal of Literary Nature and Science Writing, among other publications. She is an herbalist who teaches people how to remove invasive plants and use them to make medicines. Presently, she is working on another book about wolf-dogs in the United States. She lives on her sailboat *Whistledown* in Portland.